Introduction to Industrial Minerals

Introduction to Industrial Minerals

D.A.C. Manning

Department of Geology,
University of Manchester, UK

CHAPMAN & HALL
London · Glasgow · Weinheim · New York · Tokyo · Melbourne · Madras

Published by Chapman & Hall, 2–6 Boundary Row, London SE1 8HN, UK

Chapman & Hall, 2–6 Boundary Row, London SE1 8HN, UK

Blackie Academic & Professional, Wester Cleddens Road, Bishopbriggs, Glasgow G64 2NZ, UK

Chapman & Hall GmbH, Pappelallee 3, 69469 Weinheim, Germany

Chapman & Hall USA., One Penn Plaza, 41st Floor, New York NY 10119, USA

Chapman & Hall Japan, ITP-Japan, Kyowa Building, 3F, 2-2-1 Hirakawacho, Chiyoda-ku, Tokyo 102, Japan

Chapman & Hall Australia, Thomas Nelson Australia, 102 Dodds Street, South Melbourne, Victoria 3205, Australia

Chapman & Hall India, R. Seshadri, 32 Second Main Road, CIT East, Madras 600 035, India

First edition 1995

© 1995 D.A.C. Manning

Typeset in 10/12 Times by Photoprint, Torquay, Devon

Printed in Great Britain at the University Press, Cambridge

ISBN 0 412 55550 6

Apart from any fair dealing for the purposes of research or private study, or criticism or review, as permitted under the UK Copyright Designs and Patents Act, 1988, this publication may not be reproduced, stored, or transmitted, in any form or by any means, without the prior permission in writing of the publishers, or in the case of reprographic reproduction only in accordance with the terms of the licences issued by the Copyright Licensing Agency in the UK, or in accordance with the terms of licences issued by the appropriate Reproduction Rights Organization outside the UK. Enquiries concerning reproduction outside the terms stated here should be sent to the publishers at the London address printed on this page.

The publisher makes no representation, express or implied, with regard to the accuracy of the information contained in this book and cannot accept any legal responsibility or liability for any errors or omissions that may be made.

A catalogue record for this book is available from the British Library

Library of Congress Catalog Card Number: 94–72021

∞ Printed on permanent acid-free text paper, manufactured in accordance with ANSI/NISO Z39.48–1992 and ANSI/NISO Z39.48–1984 (Permanence of Paper).

To Gerard, Frances and Joseph

To Ursula, for ever and always

Contents

A colour plate section appears between pages 84 and 85

Preface	x
Acknowledgements	xi

1	**Introduction**	1
	1.1 Some definitions	2
	1.2 Examples of industrial minerals	6
	1.3 Place and value	8
	1.4 Industrial minerals and national economy	8
	1.5 Mineral production in industrialized countries	10
	1.6 Creation of markets through political forces	12
	1.7 Sources of information concerning industrial minerals	14
	1.8 Footnote to Chapter 1	16
2	**Aggregates for construction**	17
	2.1 What is required of an aggregate?	20
	2.2 Aggregates for tarmac and roads	21
	2.3 Aggregates for concrete	26
	2.4 Sources of aggregate within the UK	30
3	**Industrial clays: kaolin, ball clay and bentonite**	35
	3.1 Mineralogy and geology of kaolin deposits	40
	3.2 Mineralogy and geology of bentonite deposits	63
4	**Minerals for agriculture and the chemical industry**	72
	4.1 Sodium carbonate	73
	4.2 Halite and potassium salts	76
	4.3 Borates	82
	4.4 Phosphate rock	83
	4.5 Sulphur	88
	4.6 Zeolites	92

5 Fired products: the need for high temperature processing — 97
5.1 Prediction of the effects of firing — 98
5.2 The interpretation of mineralogical phase diagrams — 99
5.3 Practical application of phase diagrams — 112
5.4 Time–temperature–transformation (TTT) diagrams — 116
5.5 Notation and conventions — 118

6 Raw materials for the glass industry — 120
6.1 Glass manufacture — 122
6.2 Geology of glass raw materials — 126
6.3 Minor constituents: the use of lithium in glass manufacture — 132
6.4 Some environmental aspects of glass manufacture — 140

7 Cement and plaster — 141
7.1 Manufacture of Portland cement — 141
7.2 Setting of Portland cement — 147
7.3 Special cements — 149
7.4 Selection and blending of raw materials — 151
7.5 Plaster mineralogy and production — 155

8 Clays for construction — 159
8.1 Raw materials — 161
8.2 Mineralogical changes during firing — 166
8.3 Mineralogy of bricks — 175
8.4 Assessment of brick clay raw materials — 176
8.5 Environmental aspects of brick production — 184

9 Refractories — 185
9.1 Conditions of service — 185
9.2 Silica refractories — 186
9.3 Magnesia refractories — 189
9.4 Aluminosilicate refractories — 190
9.5 Other refractory products and raw materials — 194
9.6 Applications of refractories — 196

10 Assessment of mineral deposits — 200
10.1 Data gathering — 201
10.2 Data checking and validation — 206
10.3 Database design and creation — 209
10.4 Gridding the data — 209
10.5 Building the reserve model — 220
10.6 Model evaluation and testing — 220
10.7 Mine planning using the models — 226

11	**Disposal of waste by landfill**	227
	11.1 Use of exhausted mineral workings for waste disposal by landfill	228
	11.2 Composition of waste	229
	11.3 Landfill leachate	231
	11.4 Landfill gas	232
	11.5 Design of landfill waste disposal sites	234
	11.6 Site management	237
	11.7 Site geology and natural lining materials	241

Appendix A Reference phase diagrams 245

Appendix B Detailed quality variation for the model evaluation used in Chapter 10 251

Appendix C Computer hardware and software 260

References 262

Index 268

Preface

This book has been written to satisfy a need identified by a large number of students at Manchester for an easily accessible 'beginner's guide' to industrial minerals. Over 250 students have now taken the course which is based on the material covered here, mainly from undergraduate classes. The book has been written from the point of view of a teacher rather than an experienced industrial mineralogist, and it sets out to bridge the gap between the theoretical courses characteristic of geology degree programmes and their practical application. It is not designed to be used as a reference book but rather as an introduction for students, in which material has been selected to be illustrative rather than comprehensive. By virtue of the subject matter, the book will also seem parochial to some, but the importance of local or regional factors in many aspects of the exploitation of industrial minerals means that this approach is often appropriate. Nevertheless, the scientific principles which are discussed are applicable generally. My hope is that this book will satisfy the curiosity of some readers and encourage others to pursue further their interest in industrial minerals. If it does so, the book will have succeeded.

David Manning
Manchester

Acknowledgements

Many individuals in industry and academia have helped me over the last several years in the accumulation of material which now finds its way into this book. I would particularly like to thank George Swindle of Mentor Consultants for providing most of the material and text used for Chapter 10, and Mike Henderson (University of Manchester), John Howe (English China Clays), Gordon Witte (Watts Blake Bearne), Colin Bristow (Camborne School of Mines), John Sandford (Salvesen Brick), Gerry Bye and Phil Kerton (Blue Circle) for their comments on drafts of parts of the manuscript or for providing material. I must however remain fully responsible for the shortcomings of the book which have not been picked up prior to publication. Additional thanks are due to English China Clays and Watts Blake Bearne for the provision of material used in some of the figures. Special thanks are due to Richard Hartley for drafting most of the original figures.

Figures 3.22, 6.3, 7.5, 7.7, 8.2, 8.3, 8.4, 8.7, 11.2, 11.6, 11.7, 11.8, 11.9 and 11.10 are reproduced with the permission of the Controller of Her Majesty's Stationery Office. Figures 4.5, 6.4, 6.5, 6.6 and 6.7 are reproduced by permission of the Director, British Geological Survey: NERC copyright reserved.

The cover photograph was supplied courtesy of English China Clays.

Watts Blake Bearne and Co. plc are thanked for their financial support, which has enabled the colour plates to be included.

Introduction | 1

Human exploitation of minerals extends back for many thousands of years and, contrary to popular belief, mining may in fact be the 'oldest profession'. Early people used minerals, initially for pigments, and stone tools for grinding and cutting. We still use some of the same minerals for pigments, and although tool technology has moved on there are many parts of the world where stone tools are still used for very long-established purposes. As civilization has developed, so the exploitation of minerals has increased. Some uses are long established; for example, brick clays have been worked for over 5000 years. Although the technology of brick making has changed considerably during this time, and although clays can now be described in detail both mineralogically and chemically, brick clays are still best defined as those clays from which bricks can be made. Similarly, the basic principles of glass and cement manufacture have been known at least since Roman times, although modern techniques place increased demands on the qualities of the raw materials that are used in manufacturing processes. New uses for minerals are constantly appearing, in response to technological and, increasingly, political developments. Minerals are now used in many areas of manufacturing industry, in the development of new products and new materials. The key to understanding the role played by industrial minerals in modern society is to understand that their exploitation is led by developments in the markets. Like farmers, mineral producers only survive if they have a ready market for their produce, and they only thrive if they can husband their resources to take advantage of the opportunities that the market offers. A geologist working with industrial minerals has to recognise this truth, and is the person best qualified to use his or her knowledge of the variability of geological materials to assist in their exploitation.

1.1 SOME DEFINITIONS

A 'mineral' can be defined as a 'naturally occurring chemical compound', and so includes a wide variety of solid geological materials which represent the simplest components of rocks.

An 'industrial mineral' can be defined as a geological material (rock, mineral, liquid or gas) which is obtained by mining (in its broadest sense) and which represents a non-metallic, non-fuel raw material of commercial value. This definition is led by the proviso that the material is valuable as an industrial raw material, and so it is not governed by geological factors alone. Although it is general to exclude metal ores, coal and petroleum from the industrial minerals family, it is important to realize that these are all classed as minerals for the purposes of economics, and that they all contribute to minerals statistics when considering national or international production or trade. In addition, a more tightly constrained view of mineral definitions may inhibit imaginative exploitation (for example, consideration of the use of certain coals as a substitute for graphite in battery manufacture). However, it is general to exclude what is perhaps the most important mineral of all, namely water.

1.2 EXAMPLES OF INDUSTRIAL MINERALS

Industrial minerals include a wide variety of commodities, from lowly materials such as sand and gravel through to high value commodities such as industrial diamonds. There is however some considerable overlap between the four types of commodity: fuel, metal ore, industrial mineral and gemstone, and some minerals might have uses within a number of different applications. A number of metal ores are also used as industrial minerals:

- rutile (TiO_2) – titanium ore and source of industrial titanium oxide pigments and other chemicals
- chromite ($FeCr_2O_4$) – chromium ore and raw material for the manufacture of chromium oxide-based refractories
- bauxite – rock composed of hydrated aluminium oxides, used as aluminium ore and raw material for the refractories industry.

It is also quite common for industrial minerals to be won from an operation in which other commodities occur and might be worked (Table 1.1).

Some of these examples are metal ores (galena for lead, sphalerite for zinc, cassiterite for tin etc.), and in a number of cases the primary product derived from a mine has changed in response to changes in mineral markets. Many mines which were originally established to produce metal ores from the Mississippi Valley Type Pb–Zn deposits of the British Pennines have by now closed, and new operations have taken over, reworking old spoil heaps and mining what was previously regarded as waste, the now valued gangue minerals. The metal ores are still recovered and are not discarded as waste but are sold as by-products in an industrial minerals operation. It is also noteworthy that in a number of china clay pits

EXAMPLES OF INDUSTRIAL MINERALS

Table 1.1 Minerals won as by-products from operations for other commodities.

Industrial mineral	Coproduct/by-product	Examples
Fluorite, Baryte	galena, sphalerite	Mississippi Valley Type mineral deposits
Pyrite[1]	base metal sulphides	volcanogenic massive sulpide deposits
China clay	cassiterite deposits	granite-hosted kaolin
Fireclay	coal	opencast coal workings
Hydrogen sulphide	methane	natural gas exploitation

[1] Pyrite is used industrially as a source of sulphur in sulphuric acid manufacture (Chapter 4).

in south-west England, old underground tin mine workings are broken into as the open pits extend, but cassiterite within china clay deposits is not produced for sale.

There is a wide spectrum of industrial minerals in items which are widely taken for granted as part of the paraphernalia of everyday life. The most obvious examples are those which are used in construction, including sand, gravel and crushed rock aggregates. These materials require little or no processing. Other construction materials such as plaster, cement, bricks and glass are produced directly from industrial minerals by heating, firing or melting. Refractory bricks are needed to line the furnaces, ladles and kilns used in the manufacture of fired goods, and they in turn are produced from industrial minerals. Rather more subtle is the very widespread use of industrial minerals within consumer goods, where the mineral component might be a single ingredient within a rather sophisticated product. Examples include the use of minerals as abrasives in kitchen and bathroom cleaners, toothpaste and polish, as a filler/extender in plastics and paper, as whiteners in paint and polishes, and in household chemicals, medicines and detergents (McVey, 1994).

With such a wide range of possible applications there is considerable variation in the requirements expected of industrial minerals, in terms of physical properties and chemical purity. There is also considerable variation in the cost or monetary value of a particular industrial mineral, depending on its suitability for a particular application. Aggregates such as sand and gravel can be sold directly from the quarry for little more than the cost of excavation, and may (at 1992 prices) cost as little as £4 per tonne. Materials which require processing by firing command higher prices,

Table 1.2 Prices (per tonne) for selected industrial minerals (from *Industrial Minerals*, December 1993).

Mineral	Price	Mineral	Price
Alumina and bauxite		Gypsum	£6–12
calcined alumina, 98.5–99.5% pure	£250–310		
bauxite, abrasive grade, > 86% Al_2O_3	$95–108	Iron oxide pigments	$120–210
bauxite, refractory grade, > 86% Al_2O_3	$200–210		
		Kaolin	
Aplite, glass grade	£25	ceramic grade	£40–80
		filler grade	£50–75
Asbestos		paper coating grade	£75–120
Canadian	C$200–1750		
South African chrysotile	$200–440	Lithium minerals	
South African crocidolite	$640–920	petalite (4.2% Li_2O)	$147
		spodumene (> 7.25% Li_2O)	$385
Ball clay		glass grade spodumene (5% Li_2O)	$175
air-dried, shredded	£25–65	lithium carbonate	$4400
pulverised, air fluid	£75–115		
		Magnesite	
Baryte		raw	$45–50
white, paint grade	£180–200	calcined agricultural	£86–90
off-white	£140–150	calcined industrial (natural)	£125–270
oil drilling grade	£45–60	calcined industrial (synthetic)	£220–290
		dead-burned, general use	£130–170
Bentonite		dead-burned, for refractories	£200–280
Wyoming bentonite, foundry grade	£130–140		
Wyoming bentonite	$25–50	Mica	£100–800
soda-exchanged fuller's earth	£98–113		
civil engineering grade	£75–85	Nepheline syenite	
oil drilling grade	£80–85	glass grade	£70
		ceramic grade	£81–110

Boron minerals and compounds	
colemanite, natural (40–42% B_2O_3)	$300–365
colemanite, synthetic (42% B_2O_3)	$360–400
anhydrous borax	£820–876
decahydrate borax	£450–500
Calcium carbonate	
ground chalk, uncoated	£26–40
precipitated calcium carbonate, uncoated	£315
Chromite	
South Africa, chemical grade	$52
South Africa, foundry grade	$65
South Africa, refractory grade	$75
sand, moulding grade	£120–150
Clay	
calcined refractory clay (incl. chamotte)	£65–90
Diatomite	£325–380
Feldspar	
ceramic grade (ex-store UK)	£160
ceramic grade (ex-mine, N. America)	$56–90
glass grade (ex-mine, N. America)	$37–74
Fluorspar	
metallurgical grade	£90–95
acid grade	£150–170
Graphite	
powder	$220–440
coarse flake	$400–600
crystalline lump	$650–850
Olivine	
for blast furnace use	£9–13
refractory/foundry sand	£40–55
Phosphates	$32–49
Potash (KCl; 60% equivalent K_2O)	£70–85
Salt (NaCl)	£20–30
Silica sand	
foundry sand	£9–10
glass sand	£10–12
Sillimanite minerals	
andalusite	$170–220
kyanite	$135–170
sillimanite	£190
Slate (powder)	£50
Soda ash	$108
Sulphur	
liquid	$83–106
solid	$25–33
Talc	£135–250
Vermiculite	$65–226
Wollastonite	£275
Zircon, all grades	$190–210

reflecting energy costs and the greater expense of investing in manufacturing plant and equipment. Similarly, other materials which may need considerable refining in order to produce an industrial mineral raw material capable of reaching particular specifications for a particular application also command high prices. Although prices fluctuate with the passage of time and vary greatly depending on local conditions, some examples of the prices of industrial minerals are given in Table 1.2. This table lists some commonly traded industrial minerals, and indicates the importance of quality requirements for particular applications (which is reflected in the prices). Many industrial mineral names are unfamiliar, and most of these will be explained later in this book.

1.3 PLACE AND VALUE

The concept of 'place value' can be used to consider the way in which industrial minerals might be quarried or mined and then transported to where they are needed. This is an entirely qualitative concept, which attempts to take into account both the intrinsic monetary value of the commodity and the costs of transporting it to market. It is best illustrated with examples.

A material with a high place value is one which has a low intrinsic value and is expensive to transport; transport costs are a dominant component of the price paid by customers (Crouch, 1993). Such materials tend to be those which are relatively widely available and cheap, and tend to be extracted very close to the places in which they are needed. Examples include sand and gravel and to a lesser extent other crushed-rock aggregates. The widespread occurrence of sand and gravels suitable for aggregate use in heavily-populated southern Britain is reflected in the dense distribution of workings shown in Figure 1.1. The general ease of availability of sand and gravel close to areas where they are needed is one of the main reasons why by-product sand and gravel from china clay workings in south-west England is uneconomic to transport elsewhere in the country – the cost of transport soon outweighs the fact that the material is effectively free to any customer who will take it away.

At the opposite end of the continuum are materials with low place value. These have a high monetary value, and may be traded internationally. An example would be kaolin suitable for paper coating, which has to meet very demanding technical requirements. It commands a high price (Table 1.2), and so, although the costs of transport are significant, they are not an obstacle to international trade.

Although commodities with very low monetary value (such as sand and gravel) are widely quarried, there are fewer production centres for higher

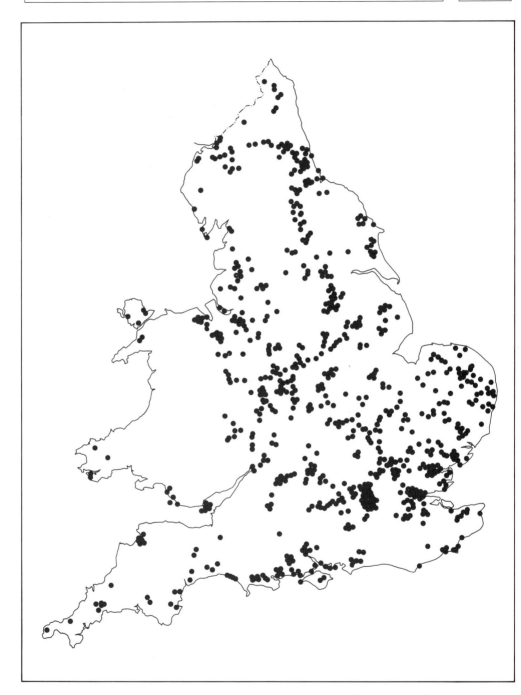

Figure 1.1 Distribution of sand and gravel workings in Britain (based on Archer, 1972).

value commodities such as limestone, gypsum and china clay. Geological controls dictate the occurrence of raw materials and their availability as resources, but transport costs can dictate whether or not extraction is viable. If a commodity occurs in the wrong place it will not necessarily be economic to work. Conversely, if a mineral deposit is particularly favourably positioned it might have economic advantages over apparently more obvious competitors. An example of this situation is provided by the development of superquarries working hard-rock aggregates on the sparsely populated coast of north west Scotland, where the only access is by sea (Whitbread and Marsay, 1992). The markets for these quarries are international, supplying the Gulf Coast states of the USA, the south-east of England and the north European coastal plain. These markets are in areas with high population density and considerable construction industry demand, but which lie on poorly-consolidated Tertiary and Quaternary sediments. Their domestic sources of hard rock aggregate may be many hundreds of kilometres inland, requiring land transport which is economically uncompetitive when compared with sea transport from coastal quarry to coastal market.

The intrinsic value of an industrial mineral depends on the development of a market for that mineral, and companies which produce industrial minerals as raw materials may be able to take an active role in stimulating new developments. An example of this is in the market for cat litter in Britain, in which fuller's earth is bagged, with little or no treatment, and then sold to domestic customers via supermarket chains. The production of cat litter *adds value* to the fuller's earth that is quarried, and the quarrying companies can obtain a higher price for the product if only because a market has been created which will stand such a price. It is also the case that competition between companies which produce and sell the same mineral commodity, for example kaolin, is in very general terms perhaps less threatening to producers than competition with other mineral products which might make adequate substitutes at lower cost. Kaolin wins the highest price when used for paper manufacture, but calcium carbonates meet some of the specifications required and can be used as a substitute in some papers, providing a possible threat to the kaolin producers. An aggressive minerals producer will constantly be on the look-out for new markets, either by creating one that is completely new, by capturing existing markets with substitute minerals or by taking over other companies which are developing substitute products.

1.4 INDUSTRIAL MINERALS AND NATIONAL ECONOMY

The development of an economy based upon manufacturing industry depends on the ability of a country to obtain mineral raw materials, either

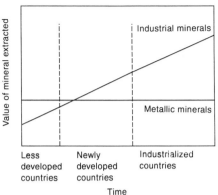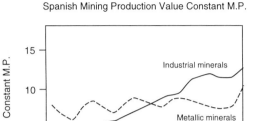

Figure 1.2 Trends in the development of industrial minerals relative to ore minerals, (a) in general and (b) for Spanish mineral production from 1960 to 1980 (from Bristow, 1987). Value expressed as millions of pesetas (M.P.).

through trade or through mining its own resources. In general terms, the importance of industrial minerals within a country's economy relates to the extent to which manufacturing industry is developed. Bristow (1987) noted a general relationship where, in less well developed countries, metallic mineral products dominate and are predominantly used for export to industrialized countries (Figure 1.2a). As industrial development proceeds the country's need for industrial minerals in support of domestic manufacturing industry increases, eventually overtaking metals in terms of their value. Bristow illustrated this general relationship with reference to changes in mineral production in Spain since the Second World War (Figure 1.2b). Although statistics are not available for a detailed consideration, this general trend also applies to Britain and other industrial countries. In the nineteenth century Britain had a major metals industry, centred on the south-west England and Pennine orefields which produced much of the metallic raw materials needed for the Industrial Revolution. Now Britain's metal ore production is almost non-existent, with minor amounts of former ore mineral products being produced as by-products in other operations and only limited mining of high value metals such as tin.

It is also valuable to look at ways in which the proportion of mineral exports changes as industrial development takes place. Mineral production in Thailand can be taken as an example. Here (Figure 1.3a) there has been growth during the last 5 years in both energy and industrial minerals, with growth and then a decline in metals. The proportions that have been exported however show a very different pattern. The proportion of metal ores that have been exported has decreased sharply, and the proportion of industrial mineral production going for export has levelled off (Figure 1.3b). This reflects the development of manufacturing industry in Thai-

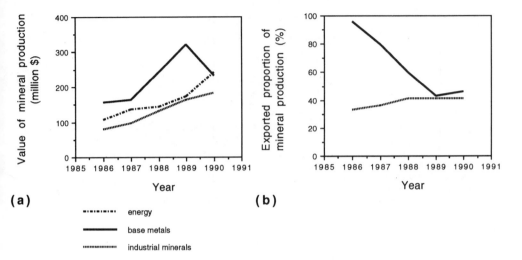

Figure 1.3 Trends in mineral production in Thailand for the period 1986–1990, showing (a) gross mineral production and (b) exports of base metals and industrial minerals (based on Intrapravich, 1992).

land, which parallels the increasingly high profile of Thai consumer goods in the supermarkets, toy and hardware shops of the developed world.

1.5 MINERAL PRODUCTION IN INDUSTRIALIZED COUNTRIES

Because of the ready availability of suitable statistical data, Britain's mineral production can be used to assess the relative value of mineral products within the economy of a developed country. Similar patterns hold for Europe and North America, bearing in mind that the occurrence of low place value commodities will cause anomalous characteristics.

At the end of the twentieth century, Britain produces a wide variety of mineral products (Figure 1.4). Fuels predominate on the basis of both value and tonnage, largely because of the contribution of North Sea oil. Figure 1.4 shows the decline of oil production and how coal production dipped during the 1984 miners' strike. Metal ore production Britain is now almost negligible.

On the basis of tonnage, over 70% of the output of industrial minerals is represented by sand, gravel, limestone, igneous rock and sandstone, most of which is used as aggregate for the construction industry and a further 15% by clays. Others make up the remaining 15%. If production of construction raw materials and clay is compared (Figure 1.5), over 90% of tonnage is accounted for by the aggregates and common clay. However, when values of minerals are compared the situation changes. Those which

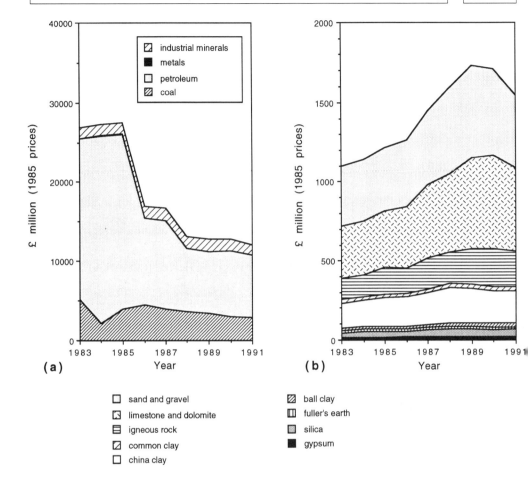

Figure 1.4 Major categories of UK mineral production: (a) divided into industrial minerals, metals, petroleum and coal and (b) divided into the dominant categories of industrial mineral (data from *UK Minerals Yearbook*, 1991, 1992).

predominantly contribute to aggregates reduce to about 80% of total production value, mostly because of the increased proportional value of china clay. This is an example of a material which has a low place value by virtue of its high intrinsic value for the paper coating industry, and it is the major industrial mineral for which Britain is a producer of international standard. The relatively high value of china clay reflects in this analysis in the distribution of employment within the minerals industry, where about 80% of employees are in the aggregates industry and over 10% in china clay (Figure 1.5c).

Having identified aggregates as the most important component of the British minerals industry in terms of volume and value, it is important to

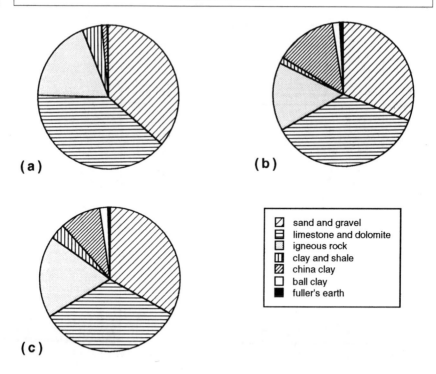

Figure 1.5 Detailed analysis of UK mineral production for clays and construction materials: proportions on the basis of (a) produced tonnage, (b) value of production and (c) employment in each minerals sector (data from *UK Minerals Yearbook*, 1991, 1992).

note that the fortunes of the aggregate industry depend entirely on activity within the construction industry. This is clearly shown by comparison of production of sand and gravel and bricks during the 1980s (Figure 1.6). In 1988 the peak of a construction boom was reached, and Figure 1.6 shows that both brick manufacture and sand and gravel extraction peaked in the same year, showing a close correlation. The production of bricks is completely independent from the mining of sand and gravel, but the consumption of the two is very closely related. Ideally, neither commodity is stockpiled, and so production is tuned to match demand as far as possible. Figure 1.6 shows the dramatic down-turn in both areas of production since 1988, which has meant that a number of brick factories, designed only to be profitable when working to capacity, have had to close.

1.6 CREATION OF MARKETS THROUGH POLITICAL FORCES

The growing level of environmental awareness and concern within Europe and North America has led to a number of changes in the use and

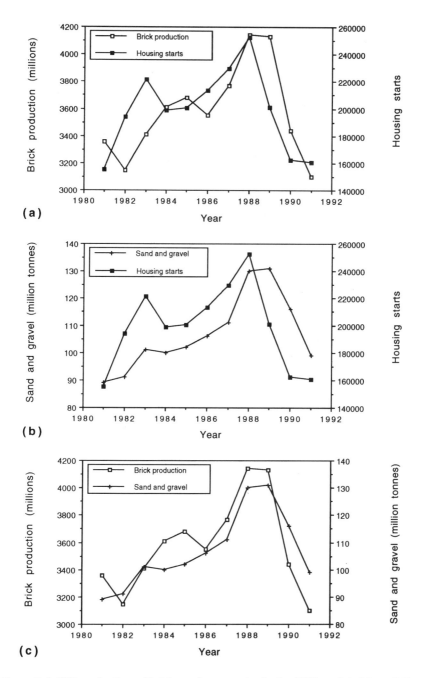

Figure 1.6 UK production of bricks and aggregates in the 1980s, related to activity in the construction industry as indicated by housing starts (data from *UK Minerals Yearbook,* 1991, 1992, and *Housing and Construction Statistics 1981–1991*, HMSO).

consumption of industrial minerals. An early casualty was asbestos, which has largely been banned on health grounds. Similarly, the use of quartz as an abrasive in sandblasting operations has now been forbidden because of the attendant risk of silicosis, and olivine has been authorized as a substitute. This has stimulated demand for olivine, creating an entirely new market as a consequence of regulation.

In more detail, we can consider the formulation of detergents. Originally, most detergent formulations were developed to include phosphates as a water softener and builder (Austin, 1984). However, their widespread use has led to increased phosphate contents within effluents, and a pollution problem. Several European countries and US states now restrict phosphate discharges (Figure 1.7), and this required that a substitute should be found for phosphates in detergents. That substitute is found within the industrial mineral zeolite family, where both synthetic and natural zeolites are available.

Zeolites carry out some of the functions of the phosphates; by absorbing calcium they act as water softeners. Detergents which contain zeolites instead of phosphates are sold as 'environmentally friendly', and come in two shades. The snow white washing powders use a synthetic zeolite, whilst those which are off-white in colour contain a natural zeolite, clinoptilolite. Both of course are ultimately derived from quarrying, with its attendant environmental implications.

1.7 SOURCES OF INFORMATION CONCERNING INDUSTRIAL MINERALS

This introduction has ranged widely in its coverage, introducing some mineral names which may be unfamiliar, and alluding to some quite sophisticated applications. The importance of industrial minerals to society has been stressed, and the breadth of the subject is immense. The chapters which follow within this book will examine, perhaps rather superficially, selected industrial minerals and products, in order to illustrate specific applications for particular minerals and particular requirements imposed by manufacturing constraints. Some geological knowledge will be assumed, so that the book can concentrate on the properties of minerals and rocks which are important in their use as industrial raw materials.

There are a number of sources of additional information which should be consulted for a more thorough study of industrial minerals. Two major works provide encyclopaedic coverage of the different industrial minerals, their geological origin and properties:

Industrial Minerals and Rocks (Lefond, 1983)
Industrial Minerals Geology and World Deposit (Harben and Bates, 1990).

SOURCES OF INFORMATION

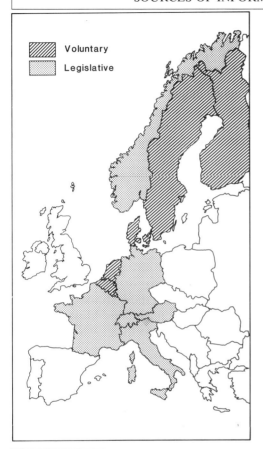

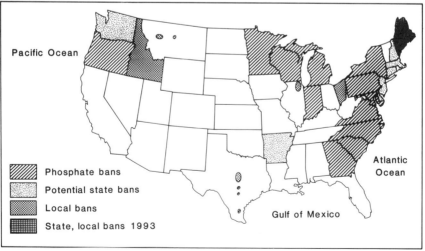

Figure 1.7 Limitations on phosphate use in detergents in (a) Europe and (b) the USA (Harries-Rees, 1992).

As suggested by the inclusion of Table 1.2, it is also essential that the monthly journal 'Industrial Minerals' (published by Metal Bulletin plc) is read on a regular basis. Other authoritative sources of information include:

United Kingdom Minerals Yearbook – compiled by the British Geological Survey, and published annually;

Mineral Commodity Summaries (and related commodity reviews) – compiled and published by the United States Department of the Interior, Bureau of Mines;

Mineral Resources Consultative Committee reports – compiled by the British Geological Survey and published in the 1970s, these deal with individual commodities in detail, but are now rather dated.

1.8 FOOTNOTE TO CHAPTER 1

As explained in this chapter, the major part of industrial mineral production is often from local sources, and international trade is limited to a wide variety but relatively small volume of commodities. Most geologists working with industrial minerals have to adjust to their own set of local conditions, and so it is inappropriate for a text such as this to attempt to generalize to global circumstances. This is the excuse for concentrating in this book on material which relates directly to mineral production in Britain, with excursions to other areas where appropriate. Readers in other parts of the world can learn from the material presented here, but, in line with their own requirements, they must personalize their knowledge of industrial minerals through consultations with local geological surveys and producers as well as the technical literature, using material cited here as a starting point.

Aggregates for construction | 2

In Chapter 1, figures drawn up using the published mineral statistics for the United Kingdom show that, in terms of tonnage, sand and gravel, limestone and dolomite and igneous rock make up 70% of Britain's mineral productivity. Within these individual categories, igneous rock and sand and gravel are almost entirely used as construction materials. Sand used as a raw material for the glass industry or for foundry purposes forms a separate category in the reported statistics. However, within the limestone and dolomite category all industrial uses are combined, and so a proportion of 'limestone' applies to glass or cement manufacture, and a proportion of 'dolomite' applies to refractories manufacture (Figure 2.1). Nevertheless, even when these alternative uses are taken into account the use of limestone and dolomite for aggregates within the construction industry is dominant.

Demand for aggregates is governed essentially by markets: first, markets need to exist (Chapter 1; Figure 1.6) and secondly, bearing in mind the high place value of these materials, markets and sources of supply need to be situated close to each other. Thus the hard rock areas of Britain have abundant deposits of aggregate materials but are relatively poorly populated and so have little demand, while the heavily developed south-east of Britain has a huge demand for aggregates but few nearby sources of supply. South and east of a line drawn from Hampshire to the Wash (Figure 2.2) there are no significant occurrences of hard rock suitable for aggregates. There are no outcrops of Carboniferous limestone, igneous rocks, metamorphic or hard clastic sediments. Thus there is considerable pressure on the closest outcrops of these, such as the Mendips for Carboniferous limestone, Charnwood Forest for granite and the West Midlands (e.g. Clee Hill) for dolerite.

Taking a broader view, heavily populated areas of the south-eastern USA and the north European coastal plains also lack convenient sources of hard rock (Figure 2.3). These markets can be supplied by material imported by sea from distant coastal quarries, such as the Glensanda

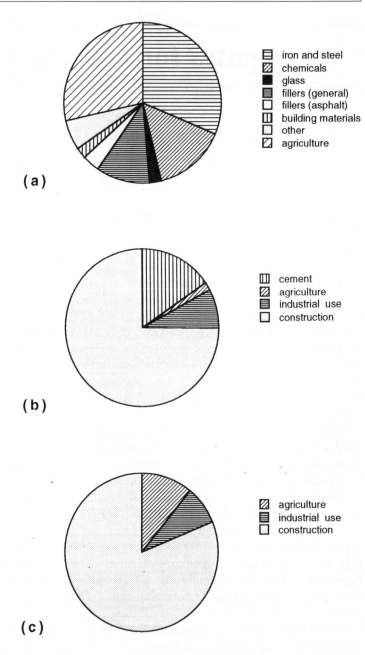

Figure 2.1 Summary of limestone and dolomite end use: (a) shows non-aggregate uses of limestone, dolomite and chalk (1991 production 14 million tonnes); (b) shows uses of limestone and chalk (1991 production 102 million tonnes); (c) shows uses of dolomite (1991 production 19 million tonnes; *UK Minerals Yearbook*, 1992)

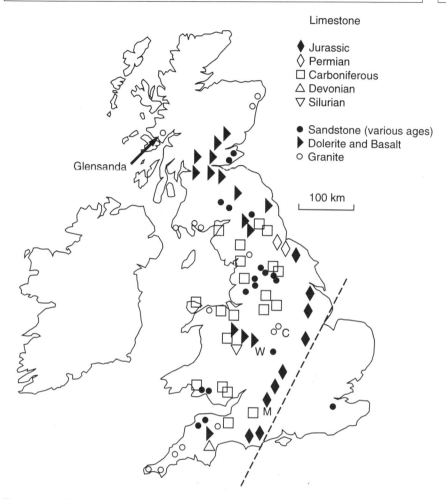

Figure 2.2 UK sources of hard rock aggregate (based on Prentice, 1990), indicating the locations of Charnwood Forest (C), the Mendips (M) and the West Midlands dolerites (W) as well as the Glensanda coastal superquarry.

operation in western Scotland (Figures 2.2 and 2.3), which exports crushed granite aggregates to Texas, north Germany and elsewhere. Other coastal quarrying operations are currently being considered for the Outer Hebrides, and areas of Spain and Norway also have potential for crushed rock exports (Whitbread and Marsay, 1992).

There is considerable environmental pressure on hard rock areas which are close to centres of population, as they are often relatively wild, when compared to nearby soft-rock agricultural land, and so are valued for recreational purposes or for their nature conservation value. They may also be in demand for up-market commuter housing, adding an irony to the

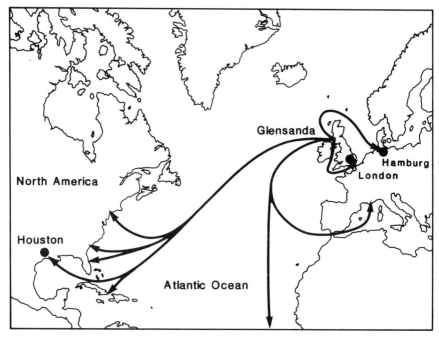

Figure 2.3 Geographical distribution of markets for aggregates quarried at Glensanda (adapted from Foster Yeoman publicity material).

fact that many who work in cities may not wish to acknowledge the requirement of those cities for raw materials. Achieving a balance between these conficting demands is far from straightforward.

2.1 WHAT IS REQUIRED OF AN AGGREGATE?

All aggregates are required to provide both bulk and strength to a material. For land fill applications (where there is no binder) strength and shape to suit the particular circumstances are most important. For aggregates which are used for tarmac or concrete (in which bitumen or cement acts as a binder to hold the aggregate together) there are a number of additional requirements which are not relevant for those used simply as land fill. In these applications, the properties of the construction material, concrete or asphalt, depend on the behaviour and properties of the aggregate just as much as on the properties of the binder. Aggregates must bond well with cement or bitumen binders and must be stable under conditions of use – they must not shrink or swell, crack or react once in use. These requirements are achieved to varying extents.

2.2 AGGREGATES FOR TARMAC AND ROADS

Ancient roads were simply made out of loose rock, perhaps specially shaped to give a regular and smooth surface, as in Roman roads. Shaped rock has been used more recently, in cobbled roads where individual cobbles might be bound together with cement or tar, and indeed moulded concrete or brick pavers are very much in fashion. Modern road construction dates back to James Macadam (1756–1836), who developed methods of binding aggregates with bitumen to produce 'tarmacadam'. The structure of a modern road is layered, to provide a foundation overlain by layers with differing properties to provide a structure which can meet the requirements of the road (Figure 2.4). Overall, the road pavement has to be sufficiently strong to take the weight of traffic that is anticipated, whilst the road surface has to resist wear and provide adequate frictional resistance to provide sufficient grip for vehicles.

Referring to Figure 2.4, the sub-base in road construction consists of crushed rock mainly 6–38 mm in size (¼"–1½"; note that the metric dimensions are 'awkward' because of conversion from imperial units), and aggregate within 0.5 m of the road surface must be frost resistant. Any material less than 4.25 mm must be non-plastic; clays are therefore undesirable, but fine grained crushed rock is acceptable. The roadbase consists of coarse aggregate bound by bitumen, and again frost resistance is important, ruling out the use of porous materials. The base course has similar requirements, but the aggregate size distribution is specified as a function of base course thickness. The wearing course is made from rolled

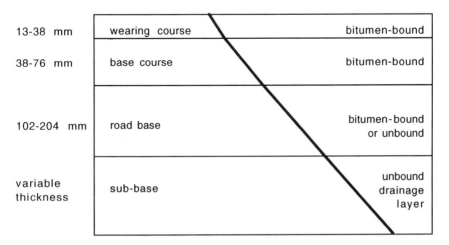

Figure 2.4 General cross-section of a road, identifying the different courses and giving their approximate thicknesses.

asphalt containing up to 30% coarse aggregate, and is covered by a layer of bitumen-coated chippings which come into contact with vehicle tyres. Aggregates used for the wearing and base courses have to meet stringent specifications, as outlined in the following section.

2.2.1 Assessment of aggregates for road construction

(a) Shape and size

Rough, poorly sorted grain-size distributions are required to maximize bonding. For unbonded materials, roundness and smoothness are important. Elimination of the <0.075 mm fraction removes the fines, especially clay minerals, which are undesirable in view of many factors (e.g. low strength, capacity to shrink or swell, plasticity, ability to lubricate surfaces and allow rock fragments to slide over one another etc.).

Aggregate particle size, which is assessed by sieving using standard sieve sizes and according to British Standard (BS) 812: 1985 is divided into coarse aggregate (>5 mm) and fine aggregate (<5 mm). The coarse aggregate fraction for road construction should meet the following size constraints:

- <38 mm (<1.5″) for the base course
- <19–25 mm (<0.75–1″) for the wearing course
- >0.075 mm

(b) Strength

Ideally, aggregate should be able to withstand the forces imposed by traffic, and strength is measured by a combination of tests, designed to simulate conditions during use: the **aggregate impact value (AIV), aggregate crushing value (ACV)** and **ten per cent fines value**, which are assessed according to guidelines laid down under BS 812: 1990:

- **aggregate impact value (AIV)** – assessing intermittent load, a standard sample in the size range 10–14 mm is subjected to 15 blows with a hammer or piston (between 13.5 and 14.1 kg in weight and falling through 381 ± 6.5 mm). The weight of fine material produced by this percussion that passes through a 2.36 mm sieve (an arbitrary size) is recorded and expressed as a percentage of the original sample weight. Low values indicate greater resistance, and should be below 20%. This test is highly reproducible, and so two values per sample are adequate, provided that they agree within to ± 1 in their numerical value.
- **aggregate crushing value (ACV)** – assessing continuous load, a sample of aggregate weighing approximately 2 kg is placed in a container and loaded to 400 kN, achieved in 10 min, using a piston. The fines produced

which pass through a 2.36 mm sieve are weighed and expressed as a percentage of the original weight. As for the AIV, two tests are carried out and should agree to within 1 percentage unit, with low numerical values indicating high resistance.
- **ten percent fines value** – this is a variation on the ACV test, designed to assess the load required to yield 10% of the material less than 2.36 mm.

(c) Mechanical durability

As well as meeting physical requirements of strength, aggregates need to be durable in service under ambient climatic conditions (especially to the effects of water and frost). Mechanical durability is assessed using the **polished stone value (PSV)**, and the **aggregate abrasion value (AAV)**.

- polish resistance – a polished aggregate provides low resistance to skids, and the propensity to develop a polish is measured according to British Standard BS 812: Part 114: 1989 using the **polished stone value (PSV)**. A sample of aggregate is polished under a rubber wheel with a load of 725 N fed with an abrasive slurry. Once polished, the sample's polish is measured using a pendulum arc friction tester which gives a coefficient of friction expressed as a percentage, which is the polished stone value. Higher values indicate greater resistance to polishing. Minimum PSV values that are acceptable depend on the traffic density and type that are expected (Table 2.1). For heavily used roads and at junctions, PSV should exceed 65. For light traffic and on straight roads with few junctions, PSV should exceed 45.

Table 2.1 Typical values for aggregate qualities for common aggregate rock types (data from Smith and Collis, 1993).

Rock type	Aggregate impact value AIV	Aggregate crushing value ACV	Aggregate abrasion value AAV	Polished stone value PSV	10% fines value (kN)
Dolerite	8	11	3.2–3.6	50–56	380
Gabbro	17	17	4.1	54	240
Granite	20	20	3.0	52	190
Hornfels	9	10	1.4	55	380
Sandstone (Pennant)	20	18	7.1	67	215
Limestone (Carboniferous)	22	23	10.0	37–39	180
Flint gravel	22	18	1.7	43	223
Quartzite gravel	15	13	1.9	53	327
Marine gravel	17	16	3.3	46	280

- **aggregate abrasion value (AAV)** – this test provides a measurement of the wear that an aggregate surface undergoes. 33 cm^3 of clean aggregate (10–14 mm) are set in resin with 6 mm protruding and placed for 500 revolutions on a rotating lap (28–30 rpm) under a 2 kg weight. Sand is fed onto the lap at a rate of 0.7–0.9 kg/min, and the loss of weight experienced during the test is expressed as a percentage of the original sample weight. The mean value for two tests is taken, to two significant figures: the lower the numerical value the more resistant the rock. Values range from 1 to 15 for acceptable materials.

Other tests of durability include the Los Angeles abrasion value (which correlates closely with the British aggregate crushing value) and the micro-Deval value (which is used widely in Europe). Both tests involve an assessment of wear in a ball mill.

(d) Physico-chemical durability

Frost resistance is important under prevailing climatic conditions in the northern hemisphere, and the ability of different aggregates to respond to volume changes associated with water ingress and/or freeze-thaw or temperature changes must be assessed. This can be done petrographically, to assess the extent and nature of weathering of the constituent minerals (to give a weathering index), or by testing resistance to wear under wet conditions (slake durability test or Washington degradation test; Smith and Collis, 1993). The **sulphate soundness test** (BS 812: Part 121: 1989 and American Society for the Testing of Materials (ASTM) C88) originated in France early in the nineteenth century, and severely tests the ability of the aggregate to withstand precipitation of magnesium sulphate within its pores (giving the **magnesium sulphate soundness value, MSSV**). A sample of aggregate within the 10–14 mm size range is subjected to five immersions in saturated magnesium sulphate solution followed by drying in an oven at 105–110°C, and the amount of material less than 10 mm in size is measured. The tests are carried out in duplicate and the mean taken as the MSSV.

Other requirements include the moisture content (which should be low and not prone to variation), and bonding ability (the aggregate should bond well with bitumen). Smooth aggregate (flint, rhyolite) does not bond well, while rough aggregates such as dolerite do. Examples of typical values for parameters which describe aggregate quality are given in Table 2.1.

In practice, few aggregates meet all the specifications that ideally are required for bitumen-bound types of construction (Table 2.1), and there are close correlations between aggregate impact value and aggregate crushing value, allowing one of these parameters alone to be used for

AGGREGATES FOR TARMAC AND ROADS

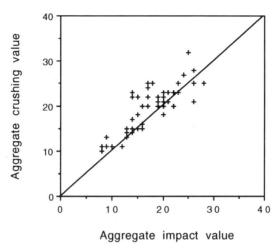

Figure 2.5 Aggregate strength parameters: correlation between aggregate impact value and aggregate crushing value, data from Smith and Collis (1993).

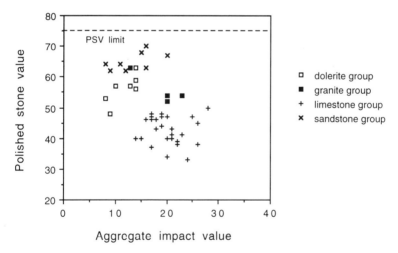

Figure 2.6 Aggregate qualities for road use: comparison of aggregate impact values and polished stone values for major crushed rock aggregate types, using data from Smith and Collis (1993). The line labelled 'PSV limit' indicates the minimum polished stone value for the most demanding applications.

comparative purposes (Figure 2.5). Figure 2.6 shows the aggregate impact value for a range of crushed rock aggregates, together with their polished stone values and limits required for particularly demanding applications. Relatively few aggregates clearly meet the most demanding requirements for polished stone value (>65 is needed for heavy traffic and 75 for the most demanding applications). Figure 2.6 shows that there is a general inverse relationship between polishing resistance and strength, in the sequence:

- sandstone group: high PSV and low AIV
- dolerite group
- granite group
- limestone group: low PSV and high AIV

In an individual quarry, measurements of mechanical properties will be made routinely and offered to customers in support of claims that a particular aggregate is suitable.

Bearing in mind the difficulties involved in selecting aggregates to meet the required specifications, a number of aggregates from different sources will be used to optimize costs. A locally available aggregate of comparatively poor quality will be used for sub-base or roadbase, with perhaps something more special for the base course. Wearing course aggregates may however be transported relatively long distances from only a limited number of quarries, particularly to meet demanding specifications for motorway or urban routes.

Coated aggregate is usually supplied directly by the quarry. However, in areas far from suitable sources, coating will be carried out on site, because coated aggregates set and become unworkable within a matter of hours. Coating plants are often observed adjacent to motorway construction sites, as well as in places where secondary aggregates (such as old railway ballast) are being reclaimed. It is also not uncommon to see piles of coated aggregate which have 'gone off', and before loading with coated aggregate, an experienced lorry driver will make sure that his tipper has a dusting of sand to prevent sticking, in case it sets before unloading.

2.3 AGGREGATES FOR CONCRETE

Concrete is made from mixtures of cement and aggregate in the ratio of approximately 1 part cement to 5 parts aggregate (both course and fine aggregate mixed in varying proportions depending on the concrete). The aggregate imparts strength and bulk, but concrete strength is measured for the finished material as well as for the raw aggregate.

For use within concrete, aggregates are routinely tested according to the mechanical specifications required for those used for road materials (particularly the ten per cent fines value and/or the aggregate impact value, according to BS 882: 1992). Important additional requirements for concrete aggregates include:

- density – heavy aggregates include most crushed rocks, with bulk densities in the range 1200–1800 kg/m^3, but light aggregates (500–1000

kg/m^3) are used in certain circumstances, such as for inner leaf walls, where suitable aggregates include clinker, pumice, slags and ashes (pulverised fuel ash or PFA), which are relatively low strength materials. Light aggregate concrete blocks are particularly valuable for sound and thermal insulation.
- grading – concrete aggregates should be poorly sorted (in the geological sense – i.e. have a wide distribution of particle size) to ensure minimum pore space. The maximum size permitted under BS 882: 1992 is 40 mm, but a nominal 20 mm coarse aggregate is widely used in Britain. Fine aggregates may be used to provide a smooth surface finish, if required.
- shape – there are no standards to govern acceptability of the shape of aggregates used in concrete, but equidimensional particles are generally preferred, and rounded particles encourage flow when pouring ready-mixed concrete into cavities or shuttering.
- surface texture – rough surfaces tend to give the best bond between the aggregate and cement, but again there are no standard requirements.
- water absorption – low water absorption is required (below 1% ideally, but acceptable below 5%) to improve resistance to frost.
- reactivity – the aggregate should not react with cement pore waters. This problem is discussed more fully below.

2.3.1 Aggregate reactivity

The reactivity of aggregates within concrete is especially important, not least because any consequent problems may only become apparent several years after construction, when the load-bearing ability of the concrete fails, necessitating demolition of the house, hospital or bridge. Some materials are particularly prone to reactivity, and should be avoided. Aggregates within a concrete are in a chemically-hostile environment, where pore fluids are alkaline, saturated with respect to calcium hydroxide from the cement paste. Of particular importance is the problem of alkali–silica reactivity.

Alkali–silica reactivity (ASR) is a phenomenon in which alkalis derived from the cement and transported by the concrete pore fluids react with siliceous aggregates to form a calcium silicate gel which adsorbs water, expanding and weakening the concrete. This reaction is poorly understood, but has immense implications for the concrete industry as it reduces the life of concrete constructions. For ASR to occur there must be a combination of three criteria: (1) presence of a reactive form of silica in excess of a critical amount, (2) sufficiently alkaline cement pore fluids, and (3) sufficient moisture. ASR can be inhibited by restriction of any or all of these criteria.

Particularly prone to ASR are aggregates which include strained quartz, volcanic glass or amorphous silica, such as flint and chert. It is partly a

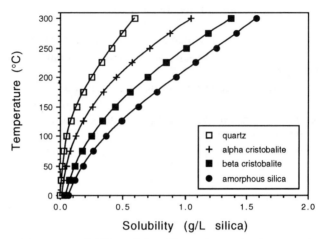

Figure 2.7 Aqueous solubility of the silica polymorphs (expressed as g/L SiO_2), calculated using formulae from Rimstidt and Barnes (1980).

consequence of the enhanced solubility of silica under alkaline conditions, and partly a consequence of the relatively high solubility of amorphous silica and strained quartz relative to unstrained quartz (Figure 2.7). In thin section (Plate B(i)), ASR is shown by the splitting and spalling of flint pebbles, and shows macroscopically by swelling or splitting/cracking of concrete surfaces and the formation of surface gels.

Although the problem of ASR has been identified and explained in terms of the formation of a secondary gel, the link with particular aggregates has not been well established, because of difficulties in isolating variables which might be significant. For example, sea-dredged aggregates are widely used in south-east England, a region where ASR is known to be a problem, and the salt water incorporated with these contributes to the alkali content of the concrete. Similarly, the alkali content of cement itself is variable, and depends on the alkali content of the shale raw materials used for cement manufacture, again affecting the properties of the finished concrete. Ways of diminishing the problem include limiting the alkali content of the concrete by varying the cement–aggregate proportions according to the alkali content of the cement, or by screening aggregates to ensure that materials suspected of being prone to ASR are excluded. Strained quartz, flint and chert might then be ruled out, creating considerable difficulties in some areas in locating suitable materials.

In addition to ASR there are a number of other reactions which are known to take place within concrete aggregates.

The *mundic problem* involves sulphides within rocks used as aggregates, and is named after the Cornish term mundic, meaning pyrite. It is associated particularly with the used of metalliferous mine waste as aggregate, a practice which has unfortunately been used historically in the production of inexpensive concrete blocks from the dumps of former mine workings. Sulphide minerals are inherently unstable, and oxidize during weathering to produce sulphuric acid, which then corrodes the concrete:

$$2FeS_2 + 7O_2 + 2H_2O = 2Fe^{2+} + 4SO_4^{2-} + 4H^+ \qquad (2.1)$$

The iron is usually oxidized to give a rusty stain on the concrete, and some additional iron may be derived from corrosion of structural steel reinforcements. It is important that sulphide-bearing aggregates are not used in concrete.

Alkali–carbonate reactivity occurs in some concretes, but is not a general problem with limestone aggregates. It seems to be restricted to aggregates in which coarse dolomite crystals are set within a finer grained clay–carbonate mixture. The mechanisms of this reaction are not fully understood, and it is not thought to be a problem in Europe.

(a) Assessment of likelihood of ASR and its prevention

A standard approach in the testing of aggregates is to make up concrete blocks using the aggregate of interest and a well characterized cement, and then to measure expansion after a period of exposure, either to external weathering or to alkaline solutions under laboratory conditions. Such tests need to be carried out over long periods, of several months or even years, and so it is not practical to screen all aggregate–cement combinations. Testing already carried out by the Building Research Establishment in the UK and other bodies has enabled a number of guidelines and procedures to be drawn up which lessen the probability of alkali–aggregate reactivity.

Petrographic investigation can be used both in the field and in the laboratory to test for the presence of the following potentially deleterious materials (Hammersley, 1989):

- porous, frost-susceptible materials
- weak or soft particles
- rock types prone to shrink–swell behaviour
- alkali-reactive forms of silica: amorphous silica or strained quartz
- alkali-reactive forms of carbonate rock
- organic matter such as coal, lignite, wood etc.
- pyrite and other sulphides
- mica
- shell fragments.

For some applications, the presence of some of these impurities can be tolerated, but for very highly specified projects, such as the Channel Tunnel, it is important to place tight controls on the quality of the aggregates to be used.

In addition to controlling the aggregates, the quality of the cement has to be considered. All cement contains alkalis, including sodium, in addition to the predominant calcium. Guidelines for the prevention of ASR, which have been suggested by the Building Research Establishment, include ensuring that the composition of the bulk concrete is such that the total sodium content of the concrete is kept below 3 kg Na_2O/m^3. This can be achieved by varying the proportion of aggregate to cement, in some cases by using pulverised fuel ash or ground blast furnace slag in addition to natural rock aggregates, and/or by using cement with a low sodium content.

2.4 SOURCES OF AGGREGATE WITHIN THE UK

There are three main sources of aggregate which can be used for construction purposes:

- sand and gravel, derived from quarries and dredging
- crushed rock, derived from hard-rock quarries
- recycled or secondary aggregates derived from demolition or refurbishment.

Photomicrographs illustrating some examples of concrete aggregates from Britain are shown in Figure 2.8.

2.4.1 Sand and gravel

Sand and gravel are widespread within Britain, most being derived from fluviatile terrace deposits such as those within the Thames and Trent valleys. Glacial sands and gravels are worked in the north, especially in Scotland. Approximately 15% of sand and gravel is derived from marine sources. The composition of these gravels depends on their sources.

- *Thames valley gravels* – in the lower reaches of the River Thames the gravels are rich in flints, derived from the Chalk, whereas in Oxfordshire Jurassic limestone fragments predominate. They are worked mainly as wet pits, in view of the high water tables adjacent to the river.
- *Trent valley gravels* – the gravels associated with the River Trent are derived mainly from pebble beds within the Triassic Sherwood Sandstone Group (Bunter pebble beds), and are rich in reworked quartz and quartzite pebbles. They are again worked mostly from wet pits adjacent to the river and its tributaries.

Gravels are also produced from other sources, including solid rock (e.g. the Sherwood Sandstone Group pebble beds which are source to the Trent valley gravels). Beaches and marine sands are worked, particularly in south-east England and the Severn estuary, by dredging where appropriate. Marine sands and gravels frequently contain shell fragments, easily recognized under the petrographic microscope (Figure 2.8a), and there are restrictions on the permitted shell content for concrete aggregates (Smith and Collis, 1993).

In view of the requirements for tarmac or concrete aggregates, gravels are not necessarily ideal or even suitable, although they are cheap to produce compared with crushed rock. They do not bond well with bitumen and develop a high polish, so they are unsuitable for roads with heavy traffic. They are, however, very suitable for concrete in view of their hardness, strength and rounded shape, but certain gravels may be prone to alkali–aggregate reactivity, depending on their content of potentially reactive phases.

2.4.2 Igneous rock

Igneous rock is suitable for both concrete and tarmac, and is of course variable depending on composition. Its distribution within Britain is essentially limited to Scotland, Wales and south-west England, with a few inliers in the English Midlands. Aggregates of good quality are produced from a limited number of quarries, tending to serve both concrete and tarmac markets. Important crushed-rock igneous aggregates include dolerite (Figure 2.8a) produced from the Carboniferous sills of the Scottish Midland Valley, Northumberland and Durham (the Whin Sill), Buxton (Derbyshire) and the Clee Hill area of the West Midlands. Crushed granite is also an important quarry product (Figure 2.8b), from locations including Shap (Cumbria), Charnwood Forest (East Midlands), Luxulyan and Hingston Down (Cornwall) and Glensanda (Argyll, Scotland).

2.4.3 Limestone and dolomite

These rocks are both comparatively soft, and so they wear quickly; they also take a good polish. They are not suitable for the most demanding road uses, but are used for roadbase and sub-base. Carboniferous limestone is however ideal for concrete in view of its lack of reactivity, as long as it is chert-free. Dolomite and chalk are less good – they are physically less strong and may be relatively permeable. Certain limestones are highly permeable, and so are unsuitable for many applications. The Carboniferous limestone is quarried extensively throughout its outcrop in Britain as a source of crushed-rock aggregate. Limestone aggregates in concrete are shown in Figures 2.8c–e.

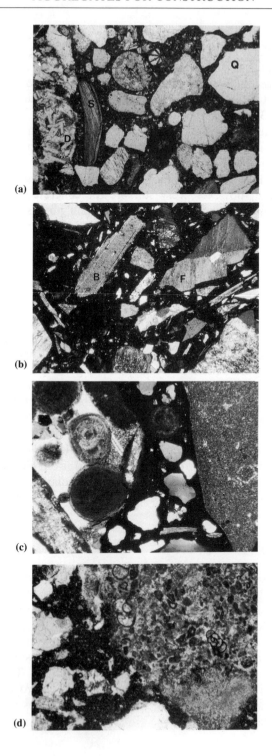

SOURCES WITHIN THE UK

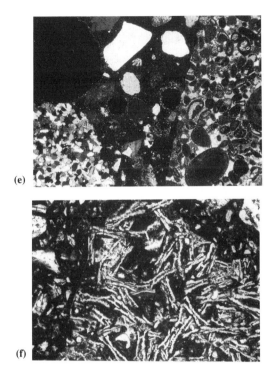

Figure 2.8 Optical photomicrographs of concrete aggregates. In all cases, the field of view is 2.85 × 1.85 mm. From a concrete produced near Newcastle-upon-Tyne, (a) shows a combination of crushed dolerite (D) as the coarse aggregate with marine sand as the fine aggregate, including a shell fragment (S) and quartz sand (Q). From a concrete produced at Shap (Cumbria), (b) shows a wide range of crushed granite fragments, in which biotite mica (B) and cloudy feldspars (F) are clearly visible. Note the lack of quartz sand; no fine aggregate other than crushed rock is used in this example. From an ornamental concrete garden product, (c) shows an oolitic limestone fragment (left) and a flint fragment (right) together with quartz sand grains in the fine fraction. Rounding of the coarse aggregate fragment shows that in this case it is derived from a gravel. (d) shows crushed Carboniferous limestone, with fossils, as the coarse aggregate and sand as the fine aggregate. (e) shows gravel-derived fragments of calcite cemented sandstone (bottom left) and oolitic limestone (right), with sand as the fine aggregate. (f) shows a fragment of blast furnace slag, with characteristic skeletal crystals.

2.4.4 Sandstones

Well-cemented sandstones satisfy the requirements of roads by being both sufficiently strong and resistant to polishing (with high PSV values), in view of their small grain size (e.g. Figure 2.8e). Examples include the Upper Carboniferous Pennant Sandstones of South Wales, and the Carboniferous and Devonian sandstones in south-west England.

2.4.5 Railway ballast

Igneous and metamorphic rocks are preferred for railway ballast in view of their high strength and abrasion resistance. There is no need to meet specifications for polishing resistance, and so rocks which are unsuitable for road use might justifiably be quarried for railway ballast and be transported relatively long distances. This provides a reservoir of aggregate close to centres of populations, where large areas of railway sidings within city centres may be redeveloped, with the possibility of recycling the ballast for use in other applications.

Further information can be obtained from the publications of the Building Research Establishment (Digests and Information Papers) and the Transport Research Laboratory (or equivalent bodies outside the UK), as well as the relevant Minerals Resources Consultative Committee dossiers *Sand and Gravel as Aggregate* (Archer, 1972), *Sandstone* (Harris, 1977a), *Igneous and Metamorphic Rock* (Harris, 1977b) and *Limestone and Dolomite* (Harris, 1982). More general information is given by Prentice (1990) and by Smith and Collis (1993), which is the definitive work on aggregates for UK-based producers and users of aggregates, containing cross-references to North American and European practices and technical tests.

Industrial clays: kaolin (china clay), ball clay and bentonite 3

Kaolin (or china clay), ball clay and bentonite are the dominant 'industrial clays', and are mined for a wide variety of uses (Table 3.1), which exploit the special properties of each of the three clay types:

- Kaolin (china clay) – is chemically inert and can be prepared as a white powder specified (in part) according to its whiteness and brightness. It is ideal for a wide range of ceramic, filling and coating applications, where the appearance of the finished product is important (Jepson, 1984). It is particularly valuable as a surface coating pigment on high quality glossy paper, which might contain up to 30% of the mineral. It is present as a filler in the paper on which this book is printed.

Table 3.1 Summary of important uses for kaolin (china clay), ball clay and bentonite

---

Kaolin (china clay)

    for paper coating and filling
    as fillers in plastics, paints and pharmaceuticals
    forms part of the ceramic body with applications similar to those of ball clay.

Ball clay

    ceramic raw material used in the production of fine tableware, stoneware (pottery), wall and floor tiles, sanitaryware and bricks
    lower grade clays increasingly are finding applications as sealing materials for landfill waste disposal sites (see Chapter 11).

Bentonite

    in the formulation of oilfield drilling fluids
    as cat litter and other animal husbandry products
    in civil engineering to produce hydraulic barriers (as in waste disposal applications; Chapter 11).

- Ball clay – is a plastic clay used in the ceramic industry to provide strength and malleability to a ceramic body prior to firing. Also, during firing, ball clay is one of the components which fuses to act as a 'cement' binding together the refractory, non-shrinking components of the ceramic body. Much of the ceramic ware produced using ball clay is moulded (such as sanitaryware) and the use of ball clay is essential to ensure that mouldings do not sag or lose shape prior to firing.
- Bentonite – is both physically and chemically reactive. It shrinks or swells in response to its ability to readily accept or release interlayer water (and organic molecules), and it exhibits important cation exchange and chemical sorption properties. It is ideal for applications where absorbance is important (particularly in the handling of waste materials), and in suspensions or slurries where its interaction with the liquid gives a fluid with particular mechanical properties (viscosity or plasticity; Odom, 1984).

In order to present acceptable qualities to the markets, all three clays undergo careful selection, testing and blending to achieve consistent properties which suit the needs of the customer. In the case of the kaolins, refining to enhance the clay content may be necessary, especially for primary kaolins (see below). Ball clays are mined selectively from individual seams, and then undergo blending and shredding; bentonites may also be treated chemically. The industrial clays can command a high price, and therefore differ from the bulk clays which are used as mined for the production of fired earthenware products such as bricks, tiles and pipes, for which the qualities of the raw materials are much less highly specified (see Chapter 8).

The contrasting and desirable properties of the three industrial clays largely arise as a consequence of their mineralogy. The dominant clay minerals are members of the kaolinite, illite and smectite groups (Figure 3.1; Table 3.2), which are present in varying proportions. The kaolins, as sold, are composed mostly of kaolinite. The clay fraction of ball clays is dominated by kaolinite, but illite is usually an essential component. The smectite clays predominate in bentonites, together with much smaller quantities of illite. The properties of the individual clay mineral groups are very strongly influenced by their crystal structure, and by the properties of the **interlayer site**. Overall, the clay minerals are characterized by layered structures in which the layers consist of sheets of aluminium and silicon covalently bonded in tetrahedral or octahedral coordination with oxygen. Individual layers are held together by electrostatic forces or by ionic bonding involving interlayer cations, which compensate for electronic charge deficiency in the aluminosilicate layers. The interlayer cations are ionically bound and in the presence of water participate in exchange reactions with ions in solution. In the smectites, water molecules also enter

INDUSTRIAL CLAYS: KAOLIN AND BENTONITE

Clay Mineral Structures

Kaolinite Group

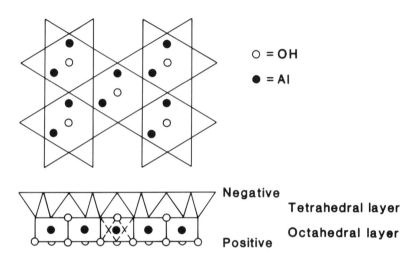

Illite Group

Vermiculite Group

Figure 3.1 Subdivision of the major clay mineral groups according to their layer structure. These 'crown and corset' diagrams emphasize the differing arrangements of the tetrahedral (crown) and octahedral (corset) components of the layers which characterize clay minerals. Note that clays belonging to the vermiculite and chlorite groups are not significant constituents of kaolins, ball clays or bentonites.

Table 3.2 Property of the kaolinite, illite and smectite clays

Property	Kaolinite group	Illites	Smectites
Structure type (tetrahedral: octahedral)	1:1	2:1	2:1
Octahedral component[1]	Dioctahedral	Mostly dioctahedral	Di- or tri-octahedral
Interlayer cations	None	K	Ca, Na
Interlayer water[2]	None (except halloysite)	None (except hydromuscovite)	Ca: 2 layers Na: 1 layer
Basal spacing	7.1 Å 10 Å in halloysite	10 Å	variable; most ~ 15 Å
Glycol absorption[3]	nil (except halloysite)	nil	takes two layers; 17 Å basal spacing
Chemical formula	$Al_4Si_4O_{10}(OH)_8$	$K_{1-1.5}Al_4(Si,Al)_8O_{20}(OH)_4$	$M^+_{0.7}(Y^{3+},Y^{2+})_{4-6}(Si,Al)_8O_{20}(OH)_4 \cdot nH_2O$
Examples	kaolinite	illite	montmorillonite (M = Na + ½Ca; Y^{3+} = 3.3Al; Y^{2+} = 0.7Mg)

[1] in dioctahedral clays, for every three octahedral sites in the layer structure two are occupied by Al or other trivalent cations and one site remains vacant. In the trioctahedral clays the octahedral sites are fully occupied by Mg, Fe or other divalent cations.
[2] gain or loss of interlayer water allows swelling or shrinkage.
[3] glycol absorption is used in the determination of clays during analysis, but also reflects their ability to absorb organic solvents.

INDUSTRIAL CLAYS: KAOLIN AND BENTONITE

the interlayer site; as dipoles, they are arranged to form a single or double structured layer (Velde, 1992). Other polar organic molecules such as ethylene glycol can similarly be accommodated in the interlayer site of the smectite family. The clay minerals are characteristically very fine grained, with crystal sizes commonly in the range 0.1–10 μm. Other minerals also occur in this size range, such as the iron and aluminium oxy-hydroxides (goethite, boehmite, diaspore and gibbsite). Although strictly these are not clay minerals, they are usually considered in mineralogical investigations of the 'clay fraction', which is the term used to describe the particle size fraction in which the clay minerals predominantly occur.

Although **kaolins** are composed predominantly of kaolinite, the raw materials may contain illite and other clay minerals as impurities. They occur either as **primary**, or **in situ**, deposits, where the clay fraction is hosted by a **matrix** of non-clay minerals at the site of its formation, or they may form **secondary** (or sedimentary) deposits, formed as a consequence of the accumulation of clays transported from locations where *in situ* deposits are eroding. The most valuable kaolins (china clays) for commercial purposes tend to yield pure white clays (partly a consequence of their chemical simplicity: Table 3.2) which can be used as a ceramic raw material in the manufacture of china or whiteware, or (very importantly) as a filler or coating for paper. Examples of major impurities which greatly reduce the value of kaolin deposits include: (1) iron oxides and hydroxides, which result in discoloured fired products, (2) smectites (or other constituents) which adversely influence the behaviour of the kaolin slurries used in high speed paper coating and (3) contaminants such as silica and fine grained feldspar which produce an abrasive slurry, which causes undue wear to process machinery.

Ball clays occur as secondary, non-marine, sedimentary deposits which may be derived from a number of primary sources. They are dominated by kaolinite, but most contain illite and occasionally traces of smectite. The non-clay fraction usually includes quartz and anatase (TiO_2), and may include iron minerals (pyrite, marcasite and siderite) and organic matter. All of these components may affect the behaviour of the clay on firing, and as far as possible their presence and content need to be known.

Bentonites are composed predominantly of clay minerals belonging to the smectite family, especially montmorillonite. There are a number of important types of bentonite:

- sodium bentonite, which is composed of sodium montmorillonite and occurs naturally as 'Wyoming bentonite'; this is a swelling bentonite.
- calcium bentonite, which is composed of calcium montmorillonite and occurs naturally as what is described in the UK as 'fuller's earth'; this is a non-swelling bentonite.

- engineered bentonite, which is a calcium bentonite that has been treated artificially using an ion exchange process to obtain a sodium bentonite, enhancing its swelling capabilities.

Bentonite clays are volcanic in origin, and represent the products of alteration of water-lain volcanic ash deposited in marine environments. In addition to the smectite clays which are present, impurities include cristobalite and zircon.

The distribution of these clay raw materials in Britain is shown in Figure 3.2, which emphasizes the concentration of ball and china clay deposits in south-west England, including Dorset, and the occurrence of bentonites within the English Jurassic outcrop. Other major clay provinces include the Westerwald (central Germany), which yields both industrial and construction clays, and the Georgia–South Carolina clay belt in the south-eastern United States (Figure 3.3).

3.1 MINERALOGY AND GEOLOGY OF KAOLIN DEPOSITS

3.1.1 Clay mineralogy of the kaolin group

The kaolinite composition is also shared by the polytypes dickite, nacrite and (with the addition of interlayer water) halloysite, which differ from kaolinite in detail in their crystal structure. All members of this clay family are dioctahedral (which means that the element in octahedral coordination is trivalent) and so only two aluminium atoms are required for every three octahedral sites (whose positive charge must total six). Dickite and kaolinite are distinguished on the basis of the arrangement of the vacancy from sheet to sheet (Figure 3.4). In kaolinite, the vacancy in the dioctahedral layer remains in the same position in each successive layer within a stack of sheets, whereas in dickite it is in the same position in each alternate layer. This rather subtle distinction leads to kaolinite exhibiting triclinic symmetry and dickite monoclinic symmetry. The two can be distinguished on the basis of careful X-ray diffraction analysis. Nacrite is chemically identical to dickite and kaolinite, but again differs in the way in which the fundamental layers are stacked, with repetition of the position of the vacancy every six layers; it is monoclinic. Kaolinite, dickite and nacrite all share a 7.1 Å basal spacing.

Halloysite has a single layer of water molecules between each aluminosilicate layer, which increases the basal spacing to approximately 10 Å. On dehydration, the basal spacing reverts to approximately 7 Å. Halloysite differs from the other kaolin group clays in that it frequently occurs as cylinders or scrolls rather than as platy grains (Figure 3.5). Relationships between halloysite and kaolinite are beautifully illustrated by Robertson and Eggleton (1991).

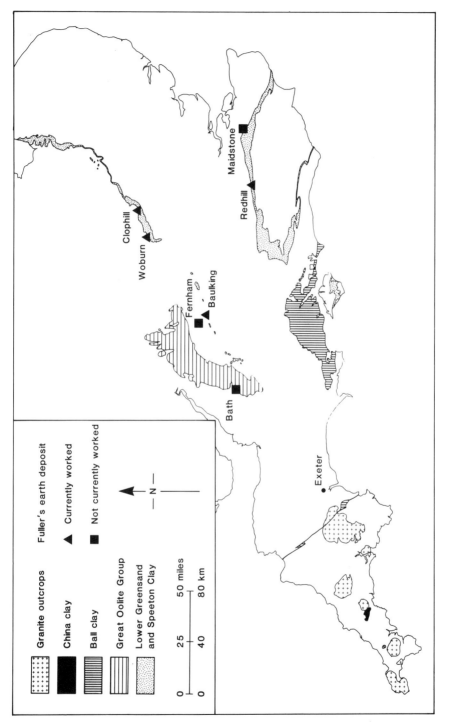

Figure 3.2 Distribution of the major industrial clay producing areas in southern England (Highley, 1975; 1984; Moorlock and Highley, 1991).

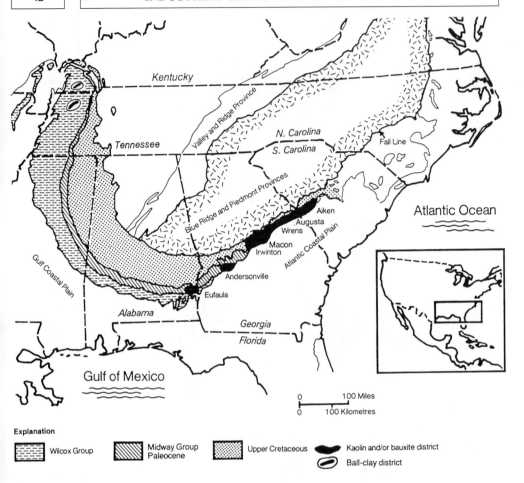

Figure 3.3 Major clay deposits of the south-eastern USA (from Harben and Bates, 1990; modified after Stack and Schnake, 1983).

3.1.2 Geology of primary (*in situ*) kaolin (china clay) deposits

In a primary deposit, the kaolin is a product of a chemical change, kaolinization, which affects a precursor phase, commonly feldspar or muscovite. Some minerals present within the rock, such as quartz and tourmaline, may resist the kaolinization process, whilst others such as biotite may decompose yielding contaminating products. In the case of the decomposition of biotite or other unstable iron-bearing silicates the products include iron oxides and hydroxides, which discolour the kaolin and reduce its value. *In situ* kaolinite deposits are commonly known as 'china clay', and the names 'kaolin' and 'kaolinite' owe their origin to the

MINERALOGY AND GEOLOGY OF KAOLIN DEPOSITS

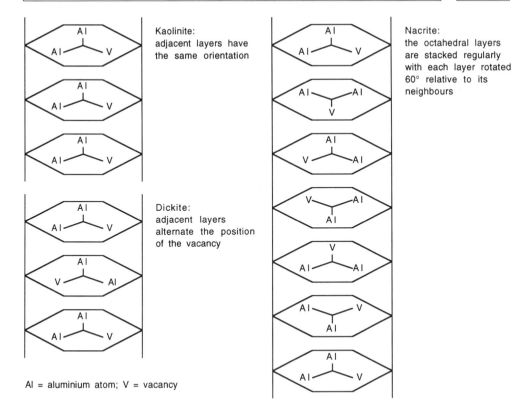

Figure 3.4 Schematic representation of the stacking arrangements for the octahedral sheets in kaolinite, dickite and nacrite.

historical importance of the deposits at Kao Lin in China; in translation, the name means 'white hill'

The process of kaolinization is at its simplest rather trivial:

$$2(Na,K)AlSi_3O_8 + H_2O + 2H^+ = Al_2Si_2O_5(OH)_4 + 4SiO_2 + 2(Na,K)^+ \quad (3.1)$$

feldspar + water → kaolinite + quartz

What is complex is determining the conditions under which this reaction takes place:

(a) Weathering

It has long been known from observation that weathering of feldspathic rocks, especially in tropical climates, yields kaolin, releasing alkalis and silica. Weathering involves meteoric fluids at low temperatures (up to 40 °C) and generally occurs close to the surface. Weathering profiles

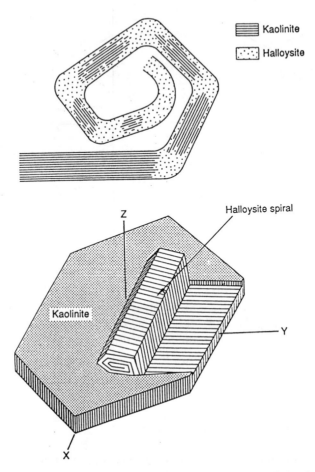

Figure 3.5 Development of halloysite coils by weathering of kaolinite sheets (from Robertson and Eggleton, 1991).

depend on the topography, climatic factors and configuration of the unsaturated zone. Deep weathering may extend to depths in excess of 100 m.

(b) Diagenesis

Considerable evidence, largely obtained from research in connection with petroleum exploration in feldspathic sandstone reservoir rocks, demonstrates that kaolin is a common diagenetic product, resulting from the alteration of feldspars during sediment burial. The diagenetic formation of kaolin again releases silica, and modifies pore fluid composition. Diagenesis starts where weathering finishes, and extends down to depths of approximately 5 km and temperatures of up to 200°C. It has been noted

MINERALOGY AND GEOLOGY OF KAOLIN DEPOSITS

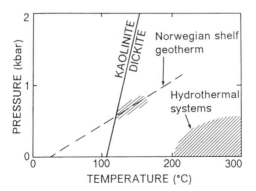

Figure 3.6 Relative thermal stability of kaolinite and dickite, showing known occurrences of dickite in North Sea reservoir rocks from the Norwegian Shelf, and in hydrothermal systems (from Ehrenberg *et al.*, 1993).

that the polytype dickite replaces kaolinite as the stable phase at a burial depth of approximately 3100–3200 m and a temperature of about 110–130°C (Figure 3.6; Ehrenberg *et al.*, 1993; cf. Macaulay, Fallick and Haszeldine, 1993). Dickite may therefore be valuable as an indicator of relatively high temperatures, but the formation of kaolinite is not necessarily confined to temperatures below 110°C. The upper temperature limit for the stability of kaolinite and its polymorphs is approximately 300°C.

(c) Hydrothermal

Hydrothermal alteration of aluminosilicate-bearing rocks is widely accepted as a mechanism for the formation of kaolinite and/or dickite. In some cases it is difficult to unambiguously dissociate the effects of weathering from those of hydrothermal alteration, because the products of high temperature hydrothermal processes are often affected subsequently by low temperature processes, including weathering. However, there is undoubtedly an important role for hydrothermal processes in 'softening up' the parent rock prior to kaolinization (Bristow, 1977). One style of kaolin occurrence which is undoubtedly associated with hydrothermal activity is **solfatara**, formed by the kaolinization of acid volcanic rocks during the waning stages of volcanic activity. The original material may have been a volcanic glass, and the kaolinized products include very fine grained silica minerals and alunite ($KAl_3(SO_4)_2(OH)_6$), the formation of which reflects the presence of sulphur of volcanic origin.

Further help in understanding the process of kaolinization is on hand in the form of information concerning the stability of the kaolinite group of

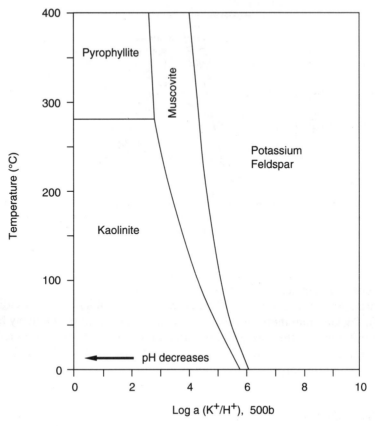

Figure 3.7 Stability of the kaolinite group as a function of temperature and fluid composition (expressed as the potassium/hydrogen ion activity ratio), calculated for a pressure of 50 MPa (500 atmospheres) using the software package GEO-CALC II: PTA-SYSTEM (Brown, Berman and Perkins, 1988).

minerals, as determined by a combination of theory and experiment, neglecting the formation of polytypes.

As discussed by Ehrenberg *et al.* (1993), the available thermodynamic data for the kaolin polytypes need to be revised before their stabilities as a function of pressure and temperature can be predicted fully. However, they note that the upper temperature limit for the kaolinite family, represented by dickite, is approximately 300°C (depending on pressure), above which temperature pyrophyllite will form. The formation of pyrophyllite at higher temperatures than those at which kaolinite is stable is reflected in the topology of the phase diagrams relating the stability of feldspar, illite, kaolinite and pyrophyllite as a function of temperature and fluid composition (e.g. Figure 3.7). These diagrams also illustrate the general relationship that the kaolinite group of minerals are stable at relatively low pH (i.e. under relatively acidic conditions), and that muscovite (as a proxy for illite, for which thermodynamic data are lacking) is formed as an intermediate phase in the alteration of the potassium feldspar.

The observation that kaolinite is unstable above approximately 300°C rules out high temperature hydrothermal origins which might be suggested by close association with greisening and tourmalinization. These hydrothermal alteration phenonema often occur within tin-mineralized granites (which often are also kaolinized), and form at 350–450°C or more.

The formation of kaolinite from alkali feldspar via muscovite (or illite) as an intermediate product can be expressed by the following reactions, in which charged species (H^+ and K^+) are in aqueous solution:

$$3KAlSi_3O_8 + 2H^+ = KAl_3Si_3O_{10}(OH)_2 + 6SiO_2 + 2K^+ \quad (3.2)$$
orthoclase muscovite quartz

$$2KAl_3Si_3O_{10}(OH)_2 + 2H^+ + 3H_2O = 3Al_2Si_2O_5(OH)_4 + 2K^+ \quad (3.3)$$
muscovite kaolinite

Greisening, represented by Reaction (3.2), takes place at temperatures in excess of the kaolinite stability limit, whilst the kaolinization Reaction (3.3) occurs below 300°C. The net result, given by adding Reactions (3.2) and (3.3), is the same as Reaction (3.1). Referring to the phase diagram shown in Figure 3.7, we can see that the transition from feldspar to kaolinite is in response to a decrease in fluid pH, matching the reactions written above.

Analogous reactions can be written involving the sodium feldspar albite and the sodium mica paragonite. However, paragonite is a rare mineral in altered feldspathic rocks, and there is some evidence from experiment that albite reacts directly to yield kaolinite, perhaps as a consequence of kinetic factors which inhibit the formation of paragonite (Manning, Gestsdóttir and Rae, 1992). This observation from the laboratory may bear on the common observation that albite (and other plagioclase species) appears to undergo kaolinization more readily than orthoclase.

Other information bearing on the origin of kaolinite in primary deposits comes from investigations of natural deposits, but may be based on indirect observations:

(a) Fluid inclusions

Fluid inclusions are samples of the fluids responsible for the crystallization of hydrothermal or diagenetic minerals which are trapped within the minerals precipitating from those fluids as growth defects. Fluid inclusions are generally microscopic (with some notable exceptions), and are necessarily studied in the minerals in which they occur. They do not occur within clay minerals, whose layer structure is not amenable to their preservation, but they are commonly observed within quartz, fluorite and other transparent minerals. The microscopic study of fluid inclusions yields an estimate of the temperature at which their host crystallized (deduced from the temperature at which multiphase inclusions homogenize on heating) and information bearing on the composition of the fluid (from the depression of freezing point). For primary kaolinite deposits, fluid inclusion data may be obtained from associated quartz, and indicate the temperature at which the quartz formed. Petrographic evidence may then allow information obtained for the quartz to be applied to the formation of the kaolinite, but very great care has to be taken in identifying quartz and kaolinite which may have formed coevally.

(b) Structural relationships

Geological investigations of primary kaolin deposits have frequently reported relationships in which the degree of kaolinization and the nature of the kaolinization are associated with major structural features, such as joints, faults or mineral veins. It is very common for kaolinized areas within granites to be associated with swarms of hydrothermal veins, which implies a genetic relationship. It has also often been observed that, in addition to this spatial association, the crystallinity of the kaolinite (a measure of the degree of order within the clay mineral structure) increases towards such features. There can be no doubt that major structural features influence the formation of kaolinite, but it may well be that their role is limited to an influence on the local permeability of the rock, which in turn influences the access of fluids responsible for weathering or low temperature hydrothermal or diagenetic kaolinization.

(c) Oxygen and hydrogen isotopes

The use of oxygen and hydrogen isotopes is a very powerful indicator of the source of the water involved in kaolinization reactions, and permits the distinction between meteoric or hydrothermal sources. Unfortunately, there are relatively few data for commercial kaolin deposits, but those which are available indicate involvement of meteoric water and so suggest a weathering origin, or isotopic re-equilibration during weathering (Sheppard, 1977).

3.1.3 Geology of secondary kaolin deposits

Secondary kaolin deposits are those which have been formed by a process of sedimentary deposition, in which a kaolinized rock (such as a granite or weathered clastic sedimentary rock) acts as a source of kaolinite. Erosion of the primary source followed by transport of the clay to a site where it can be deposited and preserved are all essential components of the overall process of formation of secondary kaolin deposits. Some secondary kaolin deposits (such as those of Georgia in the south-eastern United States) are sufficiently pure to be mined as an alternative to *in situ* china clay deposits, to produce kaolins suitable for the paper industry. As secondary kaolinite deposits, ball clays are also derived from weathered primary sources, but contain much mineral (and organic) matter other than kaolinite. The environments of deposition may be either fluviatile/lacustrine (as for the south-west England ball clays and the Cretaceous Georgian kaolins) or deltaic (as for the Tertiary Georgian kaolins). As with all sedimentary rocks, diagenetic processes may modify the mineralogy and textures of secondary kaolin deposits, in response to reactions between the pore fluids and mineral grains.

Generally, kaolinite within secondary deposits is derived from a primary source (although this might not be identified in some cases), and so ultimately is subject to the constraints described above for *in situ* alteration of feldspars and muscovite. A major concern when dealing with secondary clays is in assessing their mineralogical composition. The processes of erosion, transport and deposition may in some circumstances result in exceptionally pure deposits of kaolinite, where natural processes are the refining agents. However, it is common for other minerals derived from weathered material in the drainage catchment area, such as quartz, micas and tourmaline, to occur in addition to kaolinite. Organic matter, such as plant material introduced into the basin as well as the organic products of biological activity within the basin, may also contribute to the clay.

3.1.4 Primary and secondary kaolinite deposits in south-west England

The kaolin (china clay) and ball clay deposits of south-west England are of international importance. With a potential annual production capacity for china clay of in excess of 3 million tonnes (80–90% of which is exported), this region is second only to the United States as a kaolin producer, and lost its status as the world's leading exporter only in 1990 when the United States took first place in the league table of exporting countries. Ball clay annual production capacity is of the order of 800 000 tonnes, approximately two thirds of which is exported. The importance of south-west England as a source of kaolinite in the form of china clay and ball clay arises as a consequence of special geological factors and by its proximity to the sea, which greatly facilitates export.

The geology of south-west England is dominated by the presence of a major composite Hercynian granite batholith (Figure 3.8), which extends throughout the counties of Devon and Cornwall. Kaolin (china clay) is produced from primary deposits within the granite, particularly from the St

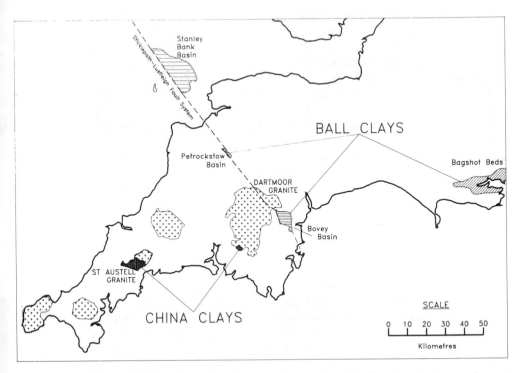

Figure 3.8 Location of primary and secondary kaolin deposits in south-west England (courtesy of Watts Blake Bearne plc). Note the occurrence offshore of the Stanley Bank Basin, which resembles structurally the onshore basins and contains ball clay sediments.

Austell pluton which yields over 80% of production (Figure 3.9). Relatively minor amounts are produced from the Dartmoor Granite, and other small deposits occur in the Land's End and Bodmin Moor Granites. Ball clays are worked in two small sedimentary basins associated with the Sticklepath–Lustleigh fault system which cuts the Dartmoor Granite – the Bovey Basin to the south and the Petrockstowe Basin to the north of this pluton. The surface area of these deposits is limited by the geometry of the basins to tens of square kilometres, but they extend to considerable depths, up to approximately 1300m. In contrast, the ball clay-bearing sequence of the Dorset area (the Lower Tertiary Bagshot Beds) is laterally much more extensive but confined to a thickness of 150m within which the ball clay deposits occur as discontinous lenses up to 6m in thickness.

Within the south-west England batholith, several granite varieties have been distinguished. Biotite granites (S-type, with an age of approximately 300 Ma) predominate, into which are intruded a later suite of tourmaline and topaz plus lithium mica bearing granites. These later intrusive phases

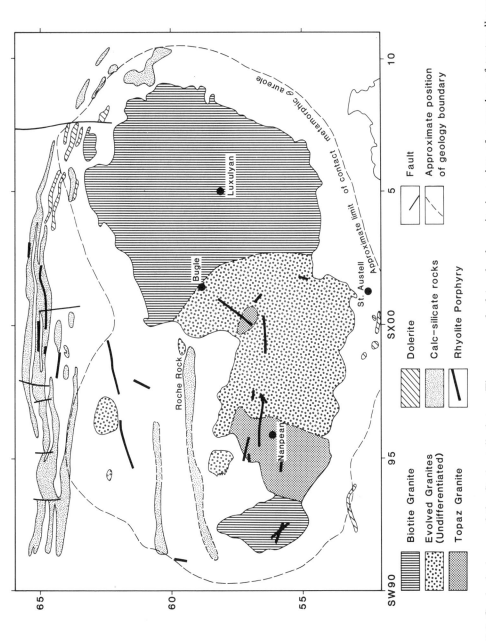

Figure 3.9 Geological map of the St Austell pluton. The area marked 'evolved granites' consists of a number of texturally varied tourmaline and lithium-mica bearing granite rocks, which show complex contact relationships.

are particularly well developed in the St Austell area (Manning and Exley, 1984), where most china clay workings are situated, and demonstrate that magmatic activity continued over a period of some 20 million years, with progressive enrichment in boron (to form tourmaline), fluorine (to form topaz) and lithium (to form lithium micas). Even before the close of magmatic activity, hydrothermal mineralization had started, with the formation of high temperature tungsten-bearing veins which are cut by late magmatic intrusions. The bulk of hydrothermal mineralization is distinctly postmagmatic, with the formation within the granites of extensive swarms of greisen-bordered tourmaline–quartz veins, some of which are mineralized with cassiterite. Later hydrothermal mineralization involves sulphide-bearing assemblages, often in large polymetallic lodes such as those worked at the Geevor and South Crofty mines. Full details of the mineralization associated with the south-west England granites are given by Jackson *et al.* (1989) and Alderton (1993), whilst the granites themselves are described fully by Floyd, Exley and Styles (1992).

Within the granite, the kaolin deposits show an overall pattern in their morphology, occurring as inverted funnels which contain kaolinized granite, with the development of kaolin apparently controlled in detail by joints (Figure 3.10).

Uneconomic kaolinization is however widespread outside the granite, as the sedimentary country rocks into which the granites were intruded show clear evidence of the development of deep weathering profiles during the Tertiary, which are dominated by clays, iron oxide/hydroxides and resistate minerals (cf. Esteoule-Choux, 1983). The kaolinized granite retains the texture of the unaltered granite, facilitating mapping, and consists of a matrix of quartz, tourmaline and micas with kaolinite taking the place of the feldspars, either completely or partially (Plate A(i)). Kaolinite also occurs within open spaces within the granite, such as shrinkage fractures (also shown in Plate A(i)) and in association with low (Plate A(ii)) or high temperature hydrothermal quartz veins. From the commercial point of view the most attractive kaolinite is that which is derived from granites which have not suffered from decomposition of iron-bearing minerals (where the resulting 'rust' stains the clay), and so biotite-bearing granites tend to be less attractive than tourmaline or lithium mica granites. Tourmaline is an iron-bearing borosilicate mineral which takes the place of biotite in magmas and hydrothermal fluids where the boron concentration is sufficiently high. Unlike biotite, it is unaffected by the process of kaolinization and remains as a resistate phase locking up iron. The lithium micas contain some iron, but less than biotite, and in any case they seem to be more stable than biotite during kaolinization.

In detail, the properties of the kaolinite are variable within the deposit. Alteration of feldspars yields large kaolinite stacks up to 100 μm long (Figure 3.11), which have a characteristic curved shape, as well as smaller

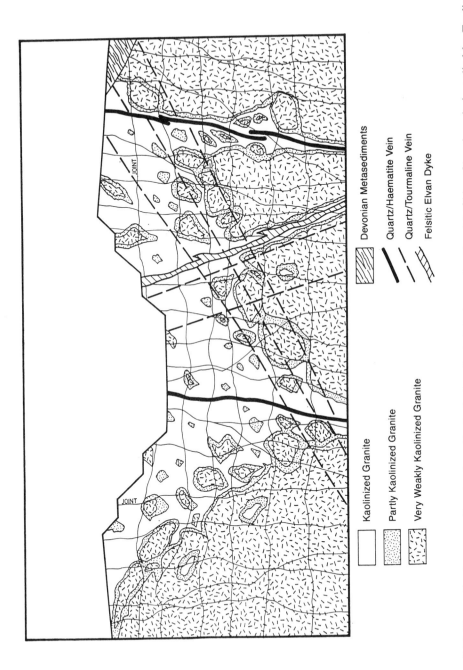

Figure 3.10 Generalized cross-section to show the typical geological features of a Cornish China Clay deposit (supplied by English China Clays International Europe plc; based on Bristow and Exley, 1994).

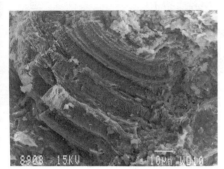

Figure 3.11

Figure 3.12

Figure 3.13

Figure 3.14

Figure 3.11 SEM photomicrograph of vermiform kaolinite stack; scale bar is 10 μm in length (supplied by English China Clays International Europe plc).

Figure 3.12 SEM photomicrograph of fine-grained kaolinite within granitic matrix; scale bar is 1 μm in length (supplied by English China Clays International Europe plc).

Figure 3.13 SEM photomicrograph of fracture-fill kaolin; scale bar is 1 μm in length (supplied by English China Clays International Europe plc).

Figure 3.14 SEM photomicrograph of silica lepispheres resting on a feldspar cleavage surface; scale bar is 10 μm in length (supplied by English China Clays International Europe plc).

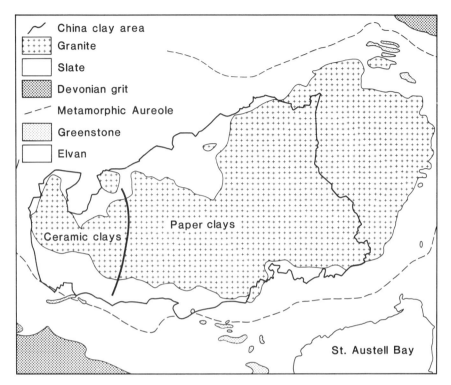

Figure 3.15 Gross distribution of clay grades in the St Austell Granite (adapted from Cornwall County Council Minerals Plan, with additional information from English China Clays International Europe plc).

individual flakes (Figure 3.12). In contrast, kaolinite from open fractures might be much finer grained (Figure 3.13). Impurities within the clay fraction include smectites as well as fine grained spherical aggregates of silica, known as lepispheres (Figure 3.14). The full extent of variability in the characteristics of the primary kaolinite deposits is not known, but there is known to be an overall geological control on clay quality, with ceramic clays associated with one rock type and paper coating clays derived from others (Figure 3.15). This is necessarily an oversimplification, for within a single pit the quality characteristics of the produced kaolins can vary tremendously.

The origin of the south-west England china clay deposits has been subject to much discussion. Arguments in favour of a hydrothermal origin have been put forward by Bray and Spooner (1983) and others, on the basis of association with hydrothermal veins and fluid inclusion evidence, but stable isotope data unambiguously identify meteoric water as the source of the oxygen and hydrogen within the clay (Sheppard, 1977). Bristow (1977) has rationalized these apparently conflicting views by

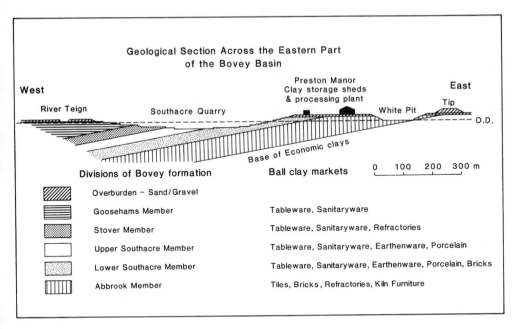

Figure 3.16 Clay sequences in the Bovey Basin, showing the gross subdivision into four members and the dominant markets which they supply (supplied by Watts Blake Bearne plc).

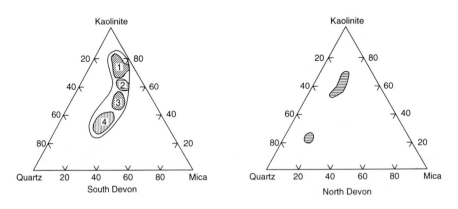

Figure 3.17 Mineralogical composition of the four different clay groups of the South Devon clays and North Devon Clays (supplied by Watts Blake Bearne plc).

proposing a two-stage origin in which hydrothermal processes 'prepared the ground' by initiating the sericitization reaction (explaining the spatial association with mineral veins) and then weathering was responsible for

Table 3.3 Chemical and mineralogical composition and physical properties (where specifications are appropriate) of kaolins (wt % except brightness, yellowness, viscosity and modulus of rupture; Bristow (1989) and Watts Blake Bearne plc company product specifications).

Component	1 Paper coating SPS	2 Filler Grade C	3 Ceramic china clay	4 Ball clay Group 1	5 Ball clay Group 4
SiO_2	47.2	47.2	47.9	48	67
TiO_2	0.04	0.14	0.03	0.9	1.4
Al_2O_3	37.6	37.4	37.2	34	22
Fe_2O_3	0.68	0.96	0.68	1.0	0.9
CaO	0.08	0.11	0.07	0.2	0.1
MgO	0.2	0.18	0.27	0.3	0.3
K_2O	1.39	1.41	1.59	1.6	2.2
Na_2O	0.08	0.07	0.08	0.2	0.3
LOI	12.7	12.5	12.3	13.8	5.8
Total	99.97	99.97	100.12	100.0	100.0
Kaolinite	93	90	88	70	34
Mica	7	9	9	16	22
Quartz	–	1	1	8	41
Feldspar	–	–	1	–	–
Carbonaceous matter	–	–	–	3.5	0
>10µm	–	5.4	2.2	–	–
<2µm	78	50	70	78–90	48–68
<0.2µm	–	–	–	16–26	14–22
ISO brightness[1]	85.6	81.0	–	–	–
Yellowness[2]	4.4	5.5	–	–	–
Viscosity concentration[3]	69.7	–	–	–	–
Fired brightness	–	–	88	68–72	45–50
Unfired modulus of rupture	–	–	14	28–49	32–42

[1] Reflectance measured at 457 nm.
[2] Difference in reflectance values measured at 570 nm and 457 nm.
[3] Percentage by weight that gives a viscosity of 5 poise at 22°C.

the bulk of the kaolinization. There can be no doubt that mineral veins within granite will locally affect the permeability of the rock allowing access for meteoric waters, and there is also no doubt that the granites were subjected to intense tropical weathering. It is much more difficult to identify processes which might have led to the formation of kaolinite before weathering occurred.

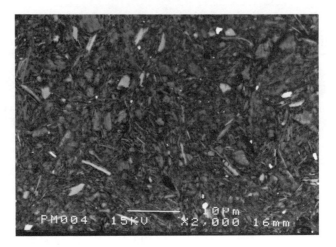

Figure 3.18 Back-scattered electron image of a polished section of ball clay. In this image, the shade of grey is determined by the composition (mean atomic number) of the mineral. Kaolinite grains show the darkest grey shade, and are irregularly oriented throughout the sample. Illite is the next whitest shade, and the bright white specks are titanium oxides.

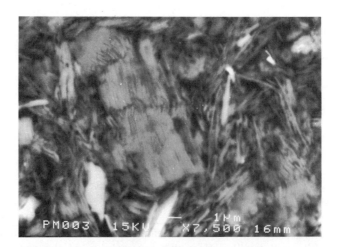

Figure 3.19 Enlargement of Figure 3.18 to show distortion of a detrital kaolinite book. Illite is also more clearly visible as whiter grains (despite poor focus, which is a consequence of the physical processes involved in generating the image, as back-scattered electrons are generated from a volume which is large compared with the scale of the features of interest).

The ball clays associated with the southwest England granites are of late Eocene–middle Oligocene age (36–28 Ma), and were deposited in two small pull-apart basins associated with the Sticklepath–Lustleigh fault system, in shallow lakes or as overbank sediments on river floodplains. The deposits are stratified, with individual beds of lignite, gravels, sands and clays ranging in thickness from a few centimetres to several metres. Within the Bovey Basin, to the south of the Dartmoor Granite, the deposit has been subdivided into four major producing sequences, the Abbrook, Southacre, Stover and Goosehams Members (Figure 3.16), within which individual seams are extracted selectively and blended to give a range of standard products, suitable for specific ceramics markets.

Mineralogically, ball clays reflect the source or sources from which they were derived. A granitic hinterland from which the transporting rivers flowed into the Bovey Basin is indicated as one source by the presence of tourmaline–quartz sands which occur within the Abbrook Member; further evidence is provided by the occurrence of zircon, cassiterite and other characteristic granite-derived minerals (Selwood *et al.*, 1984). A second source for both kaolinite and illite within the Bovey Basin is the weathered sequence of Carboniferous and Devonian sediments which surrounds the basin to the north and west. The clay-rich seams are rarely composed purely of kaolinite but can be described as mixtures of kaolinite, illite and quartz (Figure 3.17), occasionally with small quantities of smectite. On the basis of their mineralogical composition, they are subdivided into four groups (Groups 1–4 in Figure 3.17; Groups 1 and 4 in Table 3.3). The most kaolinite-rich clays occur in the Stover Member. Organic matter is frequently present, especially in the Southacre Member, where up to 50% of the lower part is composed of lignites with a variable clay content. Because of the problem of explaining how a suspension of kaolinite might settle, it is envisaged that conditions within the standing water were acidic as a consequence of a high content of organic acids (especially humic acids), and this was sufficient to flocculate the clays (Vincent, 1983). An entirely non-granitic source (or sources) is proposed for the kaolinites of the Petrockstowe Basin, which lacks tourmaline-bearing sands. In the Petrockstowe Basin the kaolinite is much finer grained and crystallographically disordered, whereas that from the Bovey Basin is relatively coarse grained and well ordered. The Petrockstowe kaolinite resembles that formed within the weathered Carboniferous sediments, whereas a component of the Bovey kaolinite resembles that developed from the granite, and appears to include the remains of kaolinite stacks that have been transported from a granitic source (Figures 3.18, 3.19). Some horizons within the Bovey Basin are derived totally or almost totally from weathered granite, whilst others contain kaolinite from both granitic and sedimentary sources and yet others contain little granite-derived material.

Although a simple, single stage sedimentary origin for ball clays appears to be acceptable, care has to be taken to identify diagenetic changes within ball clays which might have influenced the morphology or other properties of the kaolinite. For example, in the Georgia sedimentary kaolin belt, there is evidence that diagenetic processes may have involved the growth of vermicular kaolinite stacks, which are unlikely to have been transported along with the much finer kaolin particles. Such diagenetic coarsening is highly attractive to producers of kaolinite, as the coarse stacks can be delaminated and treated to produce very satisfactory paper coating clay slurries.

3.1.5 Processing and refining of china clays and ball clays

Both raw materials are worked for their kaolin content and require processing to produce blends with properties suitable for particular markets, and which optimize exploitation of variable grades.

In many locations, china clay matrix is excavated mechanically and moved by vehicle or conveyor to the point outside the pit at which wet processing starts. In south-west England, china clay is mined hydraulically using high pressure monitors: the matrix physically disaggregates and drains as a slurry to a sump in the pit, in which streams draining from different working faces can be blended. Plate A(iii) shows the contrast in whiteness of two different streams entering a sump, reflecting differing qualities in the clay that each working face can produce. From the sump, spiral or bucket wheel classifiers remove sand and gravel, and the supernatant slurry is pumped via hydrocyclones to a refining plant where size separation is continued (Figure 3.20). The characteristic grain size of the dominant minerals shows that the refining process essentially involves isolation of the clay fraction from the matrix (Figure 3.21), and attempts have been made to describe numerically particle size distribution curves for refined kaolinite concentrates (Conde-Pumpido, Ferrón and Campillo, 1988). Before being sold, the clay (which is transported and processed as a water-based suspension) is dried to have a moisture content of approximately 10%, or sold as a high solids concentration slurry. In order to produce different grades of clay, particular size fractions are isolated and sold, according to their commercial characteristics, as fillers, ceramic raw materials or paper coating pigments (Figure 3.22). To ensure adherence to specifications for particular products, clays are blended, but when the specifications cannot be achieved through blending alone additional mineral processing may be required. An example of this is chemical bleaching which is done when necessary to improve the clay whiteness. Another important property, particularly for coating pigments, is viscosity which can be adversely affected by the presence of impurities such as montmorillonite or illite, and which again can be improved through

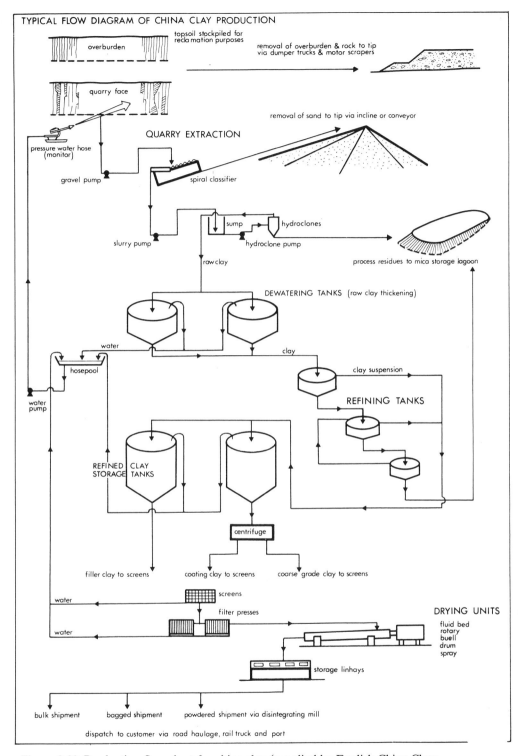

Figure 3.20 Production flow chart for china clay (supplied by English China Clays International Europe plc).

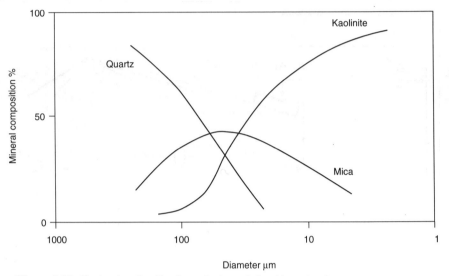

Figure 3.21 Grain size distribution of minerals in china clay (from Highley, 1984).

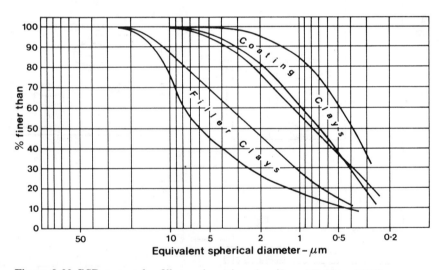

Figure 3.22 PSD curves for filler and coating clay (from Highley, 1984).

mineral processing. A selection of important properties commonly reported for china clays is given in Table 3.3.

The importance of clay particle size and shape is discussed in a series of recent papers on the subject which address applications for the use of kaolinite as a filler and coating pigment in the paper and as a filler in the plastics industries (Jennings, 1993; Slepetys and Cleland, 1993; Adams, 1993).

MINERALOGY AND GEOLOGY OF BENTONITE DEPOSITS

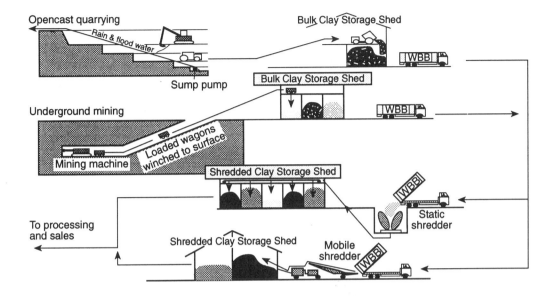

Figure 3.23 Schematic diagram of ball clay production, blending and storage (supplied by Watts Blake Bearne plc). Blending is carried out by mixing lorry-loads of clay drawn from the bulk clay storage sheds prior to the shredding process.

Ball clay is worked in opencast pits, predominantly from 3 m high benches, and there is also some very limited underground mining. An excavator selectively mines the production benches, in which individual seams have been characterized and marked before mining (Figure 3.23; Plate A(iv)). Transport from the pit is usually by dump-truck, but conveyors are also used. Subsequent processing includes shredding and blending; particular products can be derived after further refining, which can include drying, grinding or production of clay slurry or noodles. A summary of the properties reported for ball clays is given in Table 3.3: note that the titanium dioxide content is much higher for ball clays than for china clay, and this parameter needs to be specified (Chapter 10), as high values lead to fired products developing a pink colour.

3.2 MINERALOGY AND GEOLOGY OF BENTONITE DEPOSITS

Bentonites differ from china and ball clays in that they consist essentially of smectite clays, regardless of origin. The dominant smectite mineral in bentonites is montmorillonite, although other members of the smectite

Table 3.4 Compositional variation in the smectite family of minerals

Mineral	Z tetrahedral	Y octahedral	X interlayer
Dioctahedral			
Pyrophyllite	Si_8	Al_4	–
Montmorillonite	Si_8	$Al_{3.3}Mg_{0.7}$	$(½Ca,Na)_{0.7}$
Beidellite	$Si_{7.3}Al_{0.7}$	Al_4	$(½Ca,Na)_{0.7}$
Nontronite	$Si_{7.3}Al_{0.7}$	$Fe^{3+}{}_4$	$(½Ca,Na)_{0.7}$
Trioctahedral			
Talc	Si_8	Mg_6	
Saponite	$Si_{7.2}Al_{0.8}$	Mg_6	$(½Ca,Na)_{0.8}$
Hectorite	Si_8	$Mg_{5.3}Li_{0.7}$	$(½Ca,Na)_{0.7}$
Sauconite	$Si_{6.7}Al_{1.3}$	$Zn_{4-6}(Mg,Al,Fe^{3+})_{2-0}$	$(½Ca,Na)_{0.7}$

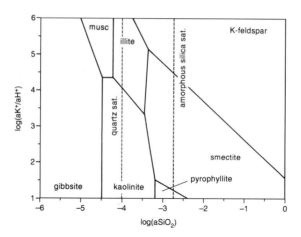

Figure 3.24 Relative stabilities of kaolinite, smectite (montmorillonite), pyrophyllite and illite as a function of fluid composition, showing fluid silica contents which correspond to saturation with respect to quartz or amorphous silica (Garrels, 1984).

group of clays may be present (Table 3.4). The smectite group of clays share the property of showing considerable shrink–swell behaviour, which is exploited in their determination by X-ray diffraction analysis. The basal spacing expands to 15 Å on saturation with Mg^{2+} cations, or to 17 Å after treatment with glycol. It decreases to 10 Å on heating to about 150°C (250°C for saponite). Note that the interlayer cation-free members of the smectite family include pyrophyllite and talc; in view of the complexity of exchanges involving the interlayer site it is often convenient to consider smectite stability by studying these two end-member species.

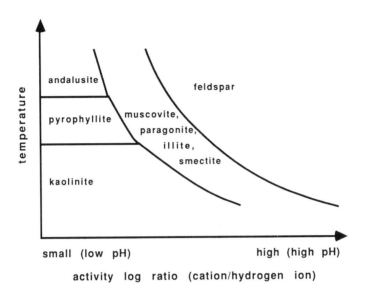

Figure 3.25 Relative stabilities of illite, kaolinite and smectite clays as a function of temperature and fluid composition. Note that the position of the field boundaries, the size of the stability fields and the nature of the phases which are stable will depend in detail on the compositions used.

Like kaolinite and illite, smectite clay stability can be expressed as a function of the composition of fluids with which it is in equilibrium using either an activity–activity diagram (Figure 3.24) or an activity–temperature diagram (Figure 3.25; cf. Figure 3.7). When constructing such diagrams it is important to state the composition of the smectite and, for simplicity, pyrophyllite is often considered as an end-member smectite composition.

Figure 3.25 very generally compares the relative stabilities of illite, kaolinite and smectite clays as a function of temperature and fluid alkali content (expressed as the appropriate ratio of K^+, Na^+ or Ca^{2+} to the hydrogen ion). This figure shows that the clays with interlayer cations are stable under conditions with higher proportions of cations to hydrogen, or conversely that kaolinite stability is favoured by low cation content relative to hydrogen. Compared with illite or smectite, kaolinite is also a low temperature phase. These phase relationships allow the origin of smectite within bentonites to be placed in a relatively well constrained physico-chemical framework.

3.2.1 Terminology

One problem of working with bentonites is that there are a number of products and materials which are described by several names, some of which

Table 3.5 Terminology of bentonites and related materials

Dominant mineral	Geographical term	Synonymous term
smectites		
Sodium montmorillonite	Wyoming bentonite (USA) Western bentonite (USA) bentonite	sodium bentonite swelling bentonite sodium-activated bentonite* sodium-exchanged bentonite* synthetic bentonite* engineered bentonite*
Calcium montmorillonite	Mississippi bentonite (USA) Southern bentonite (USA) Texas bentonite (USA) fuller's earth (UK)	calcium bentonite sub-bentonite non-swelling bentonite
Magnesium montmorillonite		saponite armargosite
Potassium montmorillonite		metabentonite potassium bentonite
Lithium montmorillonite		hectorite
non-smectites		
attapulgite[1]	fuller's earth (USA)	palygorskite
sepiolite[1]		meerschaum (lumps)
sodium sepiolite		loughlinite

* artificially produced by Na for Ca exchange of calcium montmorillonite; [1] attapulgite and sepiolite are clay minerals which have a fibrous rather than a platey morphology; their tetrahedrally coordinated aluminosilicate sheets are separated by ribbons rather than sheets of interlayer cations, permitting them to curl up. The channels provided by this structure make them excellent absorbants.

may be used differently in different countries. A summary of terminology and usage is given in Table 3.5, which includes a number of materials which are not smectite clays but which share applications with bentonites and so have acquired related names.

3.2.2 Geology of bentonite deposits

(a) In situ *deposits*

By analogy with the situation for kaolinite, *in situ* formation of smectite clays by the hydrothermal alteration of suitable parent material can yield bentonite deposits of economic significance. Such deposits are small when

Table 3.6 Examples of hydrothermal bentonites

Location	Parent rock/environment of alteration*	Primary product
Greece:		
Milos	dacite and associated tuffs/ alkaline at depth	Ca-bentonite (deep levels) kaolinite (shallow levels)
Italy:		
Sardinia	trachytic volcanic ash	Ca-bentonite
Japan:		
W. Honshu	rhyolitic ash and pumice/marine	Na-bentonite

* where known.

compared with the major sources of bentonites, which are sedimentary rocks in their own right. However, in some countries (e.g. Greece, Italy, Japan and South Africa) hydrothermal deposits are important bentonite resources.

Figures 3.24 and 3.25 show the conditions required to develop smectites as a consequence of hydrothermal alteration. Relatively high temperatures are required, depending on the fluid composition which must have cation/hydrogen ratios sufficiently high to stabilize smectite. The silica contents of the fluid must exceed quartz saturation, and to generate Ca or Na smectite the parent rock must be rich in either cation. Thus basic or especially intermediate igneous rocks represent potential parent material for hydrothermal bentonites, as the alteration of calcic plagioclase yields the required Ca or Na. A summary of parent materials for some hydrothermal bentonites is given in Table 3.6.

It is important to note that in addition to smectite, all bentonites contain other minerals which are either produced during bentonization or are residual minerals unaffected by the process. In many cases the silica mineral is cristobalite, which has a higher solubility than quartz. This helps to stabilize smectite, by constraining the fluid composition such that $\log(aSiO_2)$ is in excess of that appropriate for quartz saturation (Figure 3.24).

(b) Sedimentary bentonites

The major bentonite deposits of the world are sedimentary, and are almost always components of Mesozoic or Tertiary sequences. The most important deposits are those of Wyoming, where natural sodium montmorillonite is worked and sold as 'Wyoming Bentonite'. These deposits occur as individual beds within the Upper Cretaceous which extend for several kilometres and are up to 3 m thick. Individual beds are made up of a number of clay layers, and have sharp bases (which are often silicified) but grade upwards into overlying bentonitic shales. In addition to montmorillonite, mica (i.e. illite or muscovite), feldspar, quartz and zeolites also occur, with or without kaolinite. Iron staining may occur, but is not commercially detrimental for many applications. The Wyoming bentonites are believed to have formed from rhyolitic volcanic ash which was erupted in the Rocky Mountain area and which was altered initially as it settled through a marine water column.

In Britain, the most important naturally occurring bentonite is a calcium montmorillonite known locally as fuller's earth. There are a number of fuller's earth horizons within the Jurassic and Cretaceous, including the Jurassic horizon known stratigraphically as the Fuller's Earth Formation, which has been worked near Bath. Within the Cretaceous, fuller's earth is produced from the Lower Greensand in south-east England. Again, a volcanic source is responsible for the formation of the English fuller's earths. The likely source for the Jurassic Fuller's Earth Formation was unknown until offshore petroleum exploration identified a volcanic centre in the North Sea, of Jurassic age, and another Jurassic volcanic centre is now known to the west of Ireland. The Cretaceous deposits are believed to be sourced ultimately from volcanic centres in the Netherlands or Western Approaches.

The formation of sedimentary deposit of bentonites depends on the combination of a number of geological factors:

- there must be a source of volcanic ash
- there must be a depositional basin in which the ash can accumulate and be concentrated by natural processes of sediment sorting
- the ash must be converted by reaction with sea water to yield smectite clays
- the deposit must be protected from erosion after deposition
- there must be no further change in the clay mineral assemblage.

This last point is important in view of the known reactivity of smectite clays during sediment burial. The smectite to illite reaction is known to take place during burial of mudrocks, and has been well documented for the US Gulf Coast (Hower et al., 1976). With increasing burial, smectite is gradually transformed into mixed layer illite–smectite clays with an increasing proportion of illite layers, reaching a maximum proportion of 80% illite at a depth of approximately 3700 m. The inherent instability of smectite during burial partly accounts for the apparent lack of bentonites in Palaeozoic sequences, apart from potassium bentonites which are formed under conditions of mild metamorphism. It would be interesting to seek

MINERALOGY AND GEOLOGY OF BENTONITE DEPOSITS

bentonites in regions of Palaeozoic terrain which are known never to have suffered deep burial.

3.2.3 Uses of bentonites

The commercial value of bentonite arises from its mineralogical properties of high specific surface area, ion exchange and sorptive properties. In the assessment of bentonite deposits, it would be normal to test the prospective material according to the anticipated end use, and so to match it with a particular market. Such tests might be very specific to particular industries. An initial assessment or screening can be carried out on the basis of cation exchange capacity, or surface area measurements which can be carried out as a chemical test of the adsorption of 2-ethoxy ethanol (ethylene glycol monoethyl ether, EGME). Table 3.7 shows typical surface area measurements for selected clay minerals. There is almost an order of magnitude difference between montmorillonite and illite, allowing samples to be screened to identify those with high values which might then be selected for more detailed or more specific tests. To a first approximation, the maximum smectite content of a clay can be estimated from the surface area, which again permits evaluation during initial screening (Moorlock and Highley, 1991; Christidis and Scott, 1993).

The uses of bentonites are legion, and there are a number of processes which are carried out to treat the bentonite, once mined, to make it suitable for particular applications. It must be remembered that treatment depends on the nature of the original material. Wyoming bentonites are sodium bentonites, which swell naturally and may be immediately suitable for particular applications. In Britain, the lack of naturally occurring sodium bentonites means that treatment of calcium bentonites (the British fuller's earth) may involve sodium activation, to exchange Ca in the clay for Na and so to produce an artificial sodium bentonite. Other treatments

Table 3.7 Approximate values for the specific surface area of selected clay minerals (from Moorlock and Highley, 1991)

Mineral	Surface area (m^2/g)
Calcium montmorillonite	800
Illite	150
Kaolinite	50
Other silicate minerals	<5

Table 3.8 Examples of applications for bentonites (from O'Driscoll, 1989)

Industry	Acid activated	Untreated	Sodium activated	Organically activated
Food production	refining, decolourizing, purifying stabilizing fats and oils	–	–	–
Sulphur production	refining; bitumen extraction	–	–	–
Fire control	fire extinguisher powders	–	–	–
Pollution control	binding agents for oil on water	–	–	–
Petroleum	refining, decolourizing, purifying oils, fats, waxes; catalysts	–	–	grease thickening
Beverages/sugar	fining of wine and juices, beer stabilization; purifying saccharine juice and syrup			–
Chemical	catalysts/catalyst carriers; insecticides/fungicides; dehydrating agents; water purification; waste adsorbents; radioactive waste adsorbent			–
Paper	pigment/colour developer for copying paper; process water purification			–
Cleaning	dry-cleaning fluid regeneration	polishes; additives to washing and cleaning agents; soap production		–
Pharmaceuticals	–	earths and medicaments; creams, cosmetics (mud packs)		–
Ore production	–	binding agents for ore pelletizing		–
Construction	–	supporting suspensions for cut-off diaphragm wall construction; anti-friction agents for shaft sinking; additive to concrete and mortar		–
Waste disposal	–	sealing liners for landfill sites; composite geotextile/clay liners; radioactive waste containment		–
Ceramics	–	plasticizing of ceramic compounds; improvement of strength; fluxes		–
Horticulture, agriculture, animal care	–	soil improvement; composting; feed pellets; cat litter; liquid manure treatment		–
Drilling	–	borehole scavenging for salt water	drilling fluid formulation	
Tar	–	–	emulsification and thixotroping of tar-water emulsions, tar and asphalt coatings	
Paint/varnish	–	–	thickening, thixotroping, stabilizing and anti-setting agents	
Foundries	–	building agents for special moulding sands	binding agent for synthetic moulding stands	binding agent for anhydrous casting sands

include acid activation of the clay (where a calcium bentonite is treated with acid to improve its sorptive properties) and organic activation (where interlayer cations are exchanged for positively charged organic species). Examples of uses are given in Table 3.8, which demonstrates the versatility of this mineral.

4 | Minerals for agriculture and the chemical industry

A very wide range of minerals is exploited for use by the chemical and agricultural industries, and it would be tedious to examine a comprehensive list here. Instead, this chapter examines the geology and mineralogy of six important raw materials: sodium carbonate, halite and potash salts, borates, phosphates, sulphur and zeolites. Of particular importance are evaporite deposits, where natural processes of refining have separated soluble salts by fractional precipitation. For other materials reference should be made to the encyclopaedic texts of Lefond (1983), and Harben and Bates (1990), and for an introduction to the industrial chemical processes involved Swaddle's text (1990) is an ideal companion to this chapter although 'Shreve' (Austin, 1984) is more detailed and more widely available.

It is interesting to note that the chemical industry depends very largely on a rather limited number of bulk raw materials. Mineral raw materials are supplemented by water and atmospheric nitrogen, which is important as a source of ammonia and nitrates (especially for use in fertilizer manufacture). Of the mineral raw materials, a very small number are essential primary raw materials, and are often related by overlap between different processes (Figure 4.1). The importance of evaporites and limestones has meant, historically, that chemicals factories have ideally been sited close to natural occurrences of both minerals, especially if sources of coal or other fuels are also convenient. More specifically, the possibility of solution mining of halite (or mining sulphur as a liquid) has favoured sites immediately overlying evaporites, with import of solid limestone from nearby quarries. Many of the European centres of chemical industry, at Teesside and in north Cheshire (England), in Lorraine (France) and throughout Germany, are located according to the occurrence and mining of evaporites.

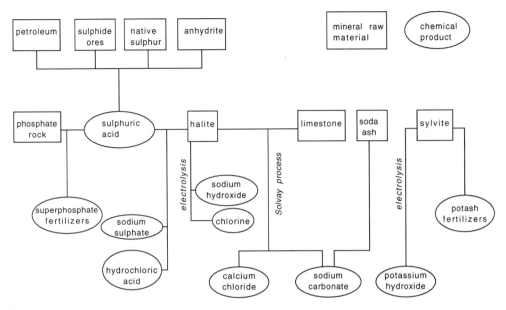

Figure 4.1 From minerals to chemicals: a summary of important raw materials and bulk products.

4.1 SODIUM CARBONATE

From 1865 until relatively recently sodium carbonate was obtained almost exclusively as an industrial product of the Solvay process, which requires limestone and halite as raw materials. However, the discovery and exploitation of natural deposits of sodium carbonate, particularly in North America, have rendered the Solvay process obsolete. It is still important in Europe, largely because there are no European sources of natural sodium carbonate, but under newly introduced conditions of free trade Solvay-produced sodium carbonate is relatively expensive and cannot compete with the imported natural material.

The Solvay process is important for historical reasons and because it is still in operation, providing a market for limestone and sodium chloride. In summary, the steps involved are as follows:

- take sodium chloride brine: $NaCl + H_2O$
- add ammonia: $NaCl + H_2O + NH_3$
- add CO_2: $NaCl + H_2O + NH_3 + CO_2 = NH_4Cl + NaHCO_3$ (4.1)
- Remove the $NaHCO_3$ precipitate by filtration, leaving a solution of ammonium chloride

- heat $NaHCO_3$: $2NaHCO_3 = Na_2CO_3 + H_2O + CO_2$ (4.2)
- calcine limestone: $CaCO_3 = CaO + CO_2$ (4.3)
- prepare hydrated lime: $CaO + H_2O = Ca(OH)_2$ (4.4)
- regenerate ammonia: $2NH_4Cl + Ca(OH)_2 = 2NH_3 + CaCl_2 + 2H_2O$ (4.5)

alternatively:

- limestone and salt go in
- sodium carbonate and calcium chloride come out
- water and ammonia are recycled.

The need for large quantities of limestone and salt accounts for the location of Solvay plants near deposits of rock salt and limestone – ICI has large factories in north Cheshire, with easy access to salt production (as brines) from Triassic evaporites near Northwich and with limestone provided by huge quarries in the Carboniferous of the Derbyshire Dome near Buxton. Similarly, in eastern France (Lorraine) the Solvay works are located near Triassic salt deposits (worked as brines) and Jurassic limestones on the eastern edge of the Paris basin. Such factories are often located in areas where salt can be mined by brine pumping, with limestone imported from surface quarries (after all, salt cannot be quarried from open pits in wet climates, whereas limestone can).

Sodium carbonate is now mostly produced from evaporitic deposits or from alkaline brines or lakes. There are a number of sodium carbonate minerals which occur within evaporites, including:

- dawsonite $NaAlCO_3(OH)_2$
- nahcolite $NaHCO_3$
- natron $Na_2CO_3 \cdot 10H_2O$
- thermonatrite $Na_2CO_3 \cdot H_2O$
- trona $Na_3H(CO_3)_2 \cdot 2H_2O$
- wegscheiderite $Na_5H_3(CO_3)_4$

These occur together with other evaporite minerals and insoluble silicate (and oxide) mineral impurities.

The largest deposits of sodium carbonate minerals are those of the Wyoming Green River Formation, with estimated reserves of 100 billion tonnes of sodium carbonate (Garrett, 1992). These immense deposits form part of an Eocene sequence which is also well known for its oil shales. The sodium carbonate mineralogy is dominated by trona, which forms beds of high purity up to 11 m in thickness, extending over areas up to 2250 km² and interbedded with halite, oil shale and other clastic sediments. There are at least 42 beds of economic interest, nearly horizontal and free from

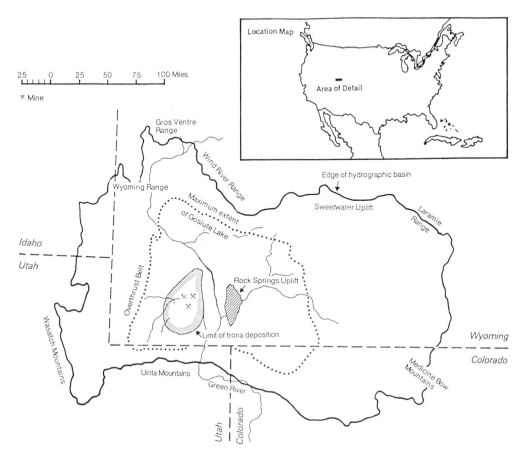

Figure 4.2 Location of Wyoming trona deposits relative to Lake Gosiute and the associated drainage basin (from Harben and Bates, 1990; adapted from Bradley, 1948, and Deardorff and Mannion, 1971).

faults. These deposits were formed on the bed of a large intra-continental alkaline lake, known as Lake Gosiute (Figure 4.2). Most of the trona beds lie directly on top of a bed of oil shale, consistent with a genetic link in which the development of alkaline conditions within saline lakes favoured prolific growth of algae. Where halite occurs it is as lenses at the top of individual trona beds. The deposits are interpreted as forming by evaporation of an alkaline lake, with sodium carbonate precipitating before halite. In the Wyoming deposits, the trona is mined using conventional underground techniques to depths of about 520 m.

Other important sources of sodium carbonate include alkaline lakes such as those of the Sierra Nevada (e.g. Searles Lake) and the African Rift Valley (e.g. Lake Magadi). These deposits are associated with alkaline springs, which in the African Rift are associated with carbonate vulcanism. Individual trona beds can reach up to 35 m in thickness, and present day formation of trona beds is well demonstrated at Lake Magadi, where annual evaporation results in layers of trona between 3–8 cm thick. Production is by mining from floating dredges or solution mining by pumping of natural or injected brines.

4.2 HALITE AND POTASSIUM SALTS

Of evaporite rocks worked to supply the chemical industry, halite and potassium salts are the most important components, often occurring together with gypsum and anhydrite. Although halite is the only mineral chloride of sodium, there are several potassium chloride minerals, and several other evaporite minerals also occur as potential ore contaminants (Table 4.1). Once they have been separated, the dominant end use for the

Table 4.1 Evaporite and associated minerals, using chemical formulae from Clark (1993).

Name	Formula	K_2O equivalent (wt %)	Status
Chlorides			
Sylvite	KCl	63	principal potash ore
Carnallite	$KMgCl_3.6H_2O$	17	potash ore and contaminant
Kainite	$KMgSO_4Cl.3H_2O$	19	important potash ore
Bischofite	$MgCl_2.6H_2O$	–	–
Halite	NaCl	–	halite ore
Sulphates			
Polyhalite	$K_2Ca_2Mg(SO_4)_4.2H_2O$	16	potash ore contaminant
Langbeinite	$K_2Mg_2(SO_4)_3$	23	important potash ore
Leonite	$K_2Mg(SO_4)_2.4H_2O$	26	potash ore contaminant
Picromerite*	$K_2Mg(SO_4)_2.6H_2O$	23	accessory
Glaserite	$K_3Na(SO_4)_2$	43	accessory
Syngenite	$K_2Ca(SO_4)_2.H_2O$	29	accessory
Blödite	$Na_2Mg(SO_4)_2.4H_2O$	–	accessory
Löweite	$Na_{12}Mg_7(SO_4)_{13}.15H_2O$	–	accessory
Vanthoffite	$Na_6Mg(SO_4)_4$	–	accessory
Kieserite	$MgSO_4.H_2O$	–	common contaminant
Hexahydrite	$MgSO_4.6H_2O$	–	accessory
Epsomite	$MgSO_4.7H_2O$	–	accessory
Anhydrite	$CaSO_4$	–	common contaminant
Gypsum	$CaSO_4.2H_2O$	–	common contaminant

* formerly known as schonite

potassium minerals is, with little further treatment, as potash fertilizers, and it is usual to express their potassium content as equivalent K_2O ('potash'), even though they may contain no oxygen. The potash fertilizers include 'muriate of potassium' (KCl), with a minimum of 60% equivalent K_2O, potassium sulphate (50% K_2O minimum) or potassium magnesium sulphate (K_2SO_4–$MgSO_4$). Other uses for potassium salts include the production of potassium hydroxide, by electrolysis of the chloride, which is then widely used in the chemicals and drugs industries. In contrast to the potassium salts, halite has no immediate application as an agrochemical, but is primarily a raw material for the chemicals industry. It is used in the Solvay process (see above), and also in the production of chlorine and caustic soda (NaOH) by electrolysis of halite brines. It is also used to produce sodium sulphate ('salt cake'), by reaction with sulphuric acid, with hydrochloric acid as a by-product.

Evaporite sequences containing potassium salts, halite and other minerals are known from many parts of the world, and characteristically show a layered structure which in many cases is deformed by halokinesis, the process by which diapirs of salt form and rise, deforming overlying rock sequences. The European Permian–Triassic Zechstein province is an example of an evaporite basin of major economic importance, extending from Poland through Germany, the Netherlands, Denmark and the North Sea to Britain. Within the Zechstein there are five main evaporite cycles (Z1–Z5), each of which shows a general sequence from clastic sediments through carbonates (limestones and dolomites; not sodium carbonates) to anhydrite, then halite and potassium salts (Taylor, 1984). There are of course deviations from this ideal sequence. Zechstein salts are extensively mined for potash in Germany, and in Britain at the Boulby mine on the north Yorkshire coast (Figure 4.3). Other European potash deposits include the Alsace deposits, which are of Oligocene age, and Tertiary deposits in Spain.

In north America, potash production is concentrated in Canada, which is the world's leading exporter and the world's second most important producer after the CIS (Figure 4.4). The potassium salts occur as a major component of the Western Canada sedimentary basin, and are of Devonian age.

Commercially exploited potassium salts occur as seams of limited thickness, generally 3–6 m, sometimes reaching 10 m. These thickness are suitable for conventional underground mining techniques, and a grade of 25–35% K_2O equivalent is commonly cited (Harben and Bates, 1990). Compared with the thicknesses of the evaporite sequences which surround them, the potassium salt seams are relatively thin. The use of conventional mining permits close control of the selection of material to be mined, and so is advantageous as it permits the exclusion of low grade ore or material

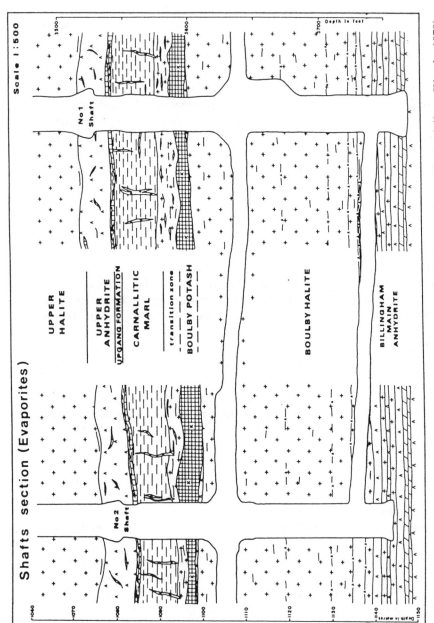

Figure 4.3 Shaft section of the potash deposits worked at Boulby mine, north-east England (from Woods, 1979).

HALITE AND POTASSIUM SALTS

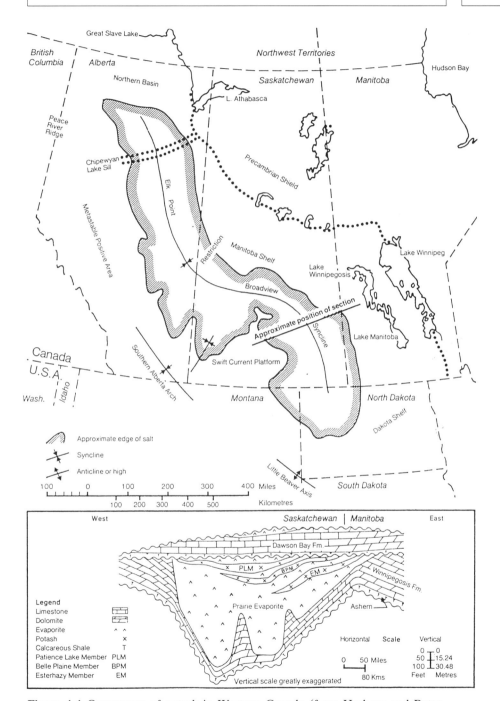

Figure 4.4 Occurrence of potash in Western Canada (from Harbern and Bates, 1990; adapted from Pearson, 1962).

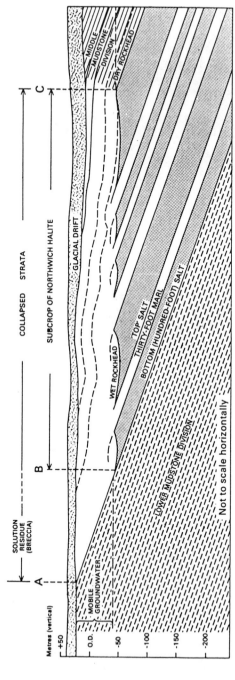

Figure 4.5 Formation of 'wet rock head' due to salt dissolution at the interface of subcropping salt-bearing strata (the Triassic Northwich Halite) and groundwater, Cheshire, UK (from Earp and Taylor, 1986).

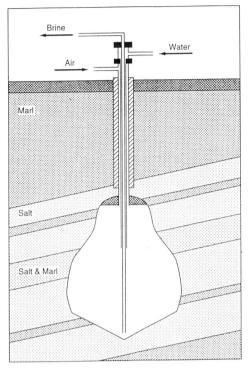

Figure 4.6 Solution mining of halite, in which compressed air and water are injected and brine withdrawn from three concentric pipes within a borehole.

rich in contaminants. The processing of potassium salts requires separation of the ore minerals from associated evaporite minerals, by means of flotation, electrostatic or solution methods.

In contrast to the potassium salts, halite occurs in sequences commonly in excess of 100 m thick in which it is the dominant evaporite mineral interbedded with clays or other classic sediments, or as domes, again with a considerable vertical extent. The major impurity is insoluble material (mostly clay) and so solution mining dominates halite production, yielding brines with NaCl contents in excess of 99.5% of the solute. Such brines have been worked for centuries as 'wild brines', produced where groundwaters intersect outcrops of halite-bearing sequences (Figure 4.5). However, wild brine production is associated with surface subsidence (as it exploits very shallow deposits), and is no longer used in areas with large populations, such as the Cheshire salt province in north-west England. Instead, underground solution mining is carried out as a controlled operation, producing from salt beds sufficiently deep to prevent surface subsidence (Figure 4.6). Mining of halite is carried out to produce solid rock salt, mainly for use on roads during winter to prevent icing. In

4.3 BORATES

In addition to sodium and potassium salts, evaporites (and associated brines) are the most important source of boron compounds. Boron occurs naturally as borate minerals, especially borax, and is important as a raw material for the glass industry (to produce glass fibre insulation and glass able to withstand thermal shock, such as 'Pyrex'). It is also used in cleaning agents, fire retardants, agriculture and metallurgy. Historically, one of the earliest uses for borax was in the preparation of mummies.

The most important boron minerals which occur naturally are listed in Table 4.2; they consist of a limited number of compositions, which have differing water contents. Like gypsum and anhydrite, the composition which is stable is governed by the degree of diagenesis or weathering and the amount of water which is available. They are therefore very susceptible to dehydration and rehydration reactions which may take place as a consequence of burial and structural deformation, and can show complex relationships between the distribution of differing mineral assemblages and structural features. Associated minerals include the arsenic sulphides realgar (AsS) and orpiment (As_2S_3). The ultimate source of the boron (and arsenic) is thought to be as a component of gaseous volcanic emissions, which affect the composition of evaporitic water bodies with which they interact, allowing the sodium and calcium borates listed in Table 4.2 to precipitate.

There are two dominant producers of boron minerals, the western United States (California: such as the place named Boron, Death Valley and Searles Lake) and Turkey. Argentina, Chile, Peru, China and the former Soviet Union also produce significant amounts. The Californian deposits consist of contrasting styles of mineralization. At Boron, borax, kernite, colemanite and ulexite occur as components of a Miocene sedimentary sequence formed in an evaporitic lake fed by sodium borate-bearing spring waters. Similarly, the Death Valley deposits consist of colemanite- and ulexite-bearing lake deposits. The solubility of the borate minerals is exploited during processing, by dissolution into aqueous liquors from which borax (decahydrate or pentahydrate) is precipitated. At Searles Lake, boron-bearing brines are worked directly.

The deposits of Turkey are dominated generally by the calcium borate mineral (ulexite and colemanite), with sodium borate such as borax occurring only at Kirka (Inan *et al.*, 1973). The deposits are all hosted by lacustrine sediments, interbedded with volcanics, of Tertiary age. Turkey is the world's leading exporter of boron minerals.

Table 4.2 Borate minerals, using chemical formulae from Clark (1993).

Mineral	Chemical Formula	wt % B_2O_3	Comments
Borax (tincal)	$Na_2B_4O_5(OH)_4.8H_2O$	36.5	dominant ore
Tincalconite	$Na_2B_4O_7.5H_2O$	47.8	accessory
Kernite	$Na_2B_4O_6(OH)_2.3H_2O$	51.0	important ore, often hydrated to borax
Ulexite	$NaCaB_5O_6(OH)_6.5H_2O$	43.0	important ore
Probertite	$NaCaB_5O_7(OH)_4.3H_2O$	49.6	secondary/accessory
Inyoite	$CaB_3O_3(OH)_5.4H_2O$	37.6	minor ore mineral
Priceite	$Ca_4B_{10}O_{19}.7H_2O$	49.8	ore (Turkey)
Meyerhoffite	$Ca_2B_6O_{11}.7H_2O$	46.7	accessory
Colemanite	$Ca_2B_6O_{11}.5H_2O$	50.8	important ore
Sassolite	H_3BO_3	56.4	natural boric acid; of historical interest
Hydroboracite	$CaMgB_6O_{11}.6H_2O$	50.5	–
Szaibelyite	$MgBO_2(OH)$	41.4	ore (Russia)
Boracite	$Mg_3B_7O_{13}.Cl$	62.2	can occur in potash deposits
Howlite	$Ca_2B_5SiO_9(OH)_5$	44.5	accessory
Kurnakovite	$MgB_3O_3(OH)_5.5H_2O$	37.3	accessory

4.4 PHOSPHATE ROCK

Phosphates, like potash, are essential raw materials for the fertilizer industry, with annual world production of phosphate rock of in excess of 150 million tonnes of which some 90% is used for fertilizers. The dominant phosphate mineral is apatite, whose formula is most simply expressed as $Ca_5(PO_4)_3(OH)$. Within this formula, fluorine and chlorine can substitute for the hydroxy group, and carbonate can substitute for phosphate, with charge balance maintained by the introduction of an additional fluorine (or hydroxyl). Apatite is a common component of both igneous rocks, where it tends to be a hydroxy–fluor apatite $(Ca_5(PO_4)_3(OH,F))$, and sedimentary rocks, where it is usually a carbonate–fluor apatite known as francolite $(Ca_5(PO_4,CO_3,OH)_3(OH,F))$.

The phosphorus content of an ore is expressed as wt% P_2O_5, with a hypothetical maximum value of 42% for fluorapatite. Phosphate rock normally requires a P_2O_5 content in excess of 20% to be considered economically viable. Impurities include associated silicate (e.g. quartz, chert, feldspar, clay and mica) and carbonate (calcite, dolomite) minerals, which are reduced or removed by beneficiation involving crushing, sizing and flotation to give a P_2O_5 content of 27–40%. The resulting concentrate is dissolved in sulphuric acid to produce superphosphate which is a mixture of typically 32% $CaHPO_4 + Ca(H_2PO_4)_2 \cdot H_2O$ and 50% $CaSO_4$ with minor amounts (3%) of adsorbed phosphoric acid (H_3PO_4), with a P_2O_5 content

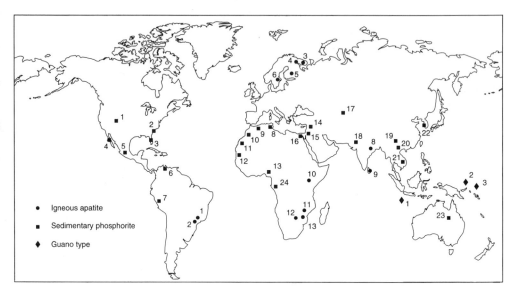

Figure 4.7 Global distribution of phosphate deposits (Table 4.3; compiled from Cook, 1984 and Harben and Bates, 1990).

of up to 20%. In producing this mixture the insoluble apatite is converted into acid calcium phosphates, which are soluble and so able to enter soil solutions for uptake by plants. Further treatment with sulphuric and/or phosphoric acid (itself produced from phosphate rock) yields products with increasingly concentrated soluble phosphate, such as concentrated superphosphate ($Ca(H_2PO_4)_2 \cdot H_2O$) with a P_2O_5 content of 56%.

Although apatite is a common accessory mineral in many igneous, sedimentary or metamorphic rocks, it is limited in its occurrence in quantities sufficient to be mined economically. Some 85% of world production is derived from sedimentary phosphate rocks, with the bulk of the remainder derived from igneous rocks. Guano, containing the phosphate minerals brushite (($CaHPO_4) \cdot 2H_2O$) and monetite ($CaHPO_4$), is no longer an important source of phosphates.

Sedimentary phosphates are predominantly marine rocks, in which the formation of phosphate minerals relates to the biological productivity of the oceans. At present, phosphate deposits are formed as concretionary horizons along the west coasts of certain continents, in response to the upwelling of phosphate-rich cold water from the deep ocean. Examples include those of Namibia and Morocco (Figure 4.7 and Table 4.3), although it is possible that some of these phosphates may be relict, bearing no relationship to current patterns of ocean circulation (Cook, 1984). For the formation of a phosphate deposit, the upwelling current must enter a shallow water region of high organic productivity, with little terrigenous

Plate A

(i) Kaolinized granite cut by kaolinite-filled en-echelon veins (St Austell, Cornwall). Note that the original magmatic texture of the granite is preserved in the matrix to the veins. The lens cap is 6 cm in diameter.

(ii) Intense kaolinization symmetrically developed adjacent to a 2 m thick quartz vein (left of centre), with weakly kaolinized granite to the right and (outside the field of view) to the left of the kaolinized zone (St Austell, Cornwall).

(iii) Streams of kaolin slurry converging at the sump in the bottom of a china clay pit (St Austell, Cornwall). Note the subtle difference in colour of the two streams, which reflects differing clay quality in the two working faces from which the streams are derived. The structures to the right of the field of view contain the pumps which take the slurry out of the pit. (Photograph supplied by P.I.Hill)

(iv) Beds of ball clay showing variation in their content of lignitic material, cut by a fault (Bovey Basin, Devon). Note the shadow cast by the flags which are used to identify individual seams for selective mining (at the level of the student's hand and head).

Plate B

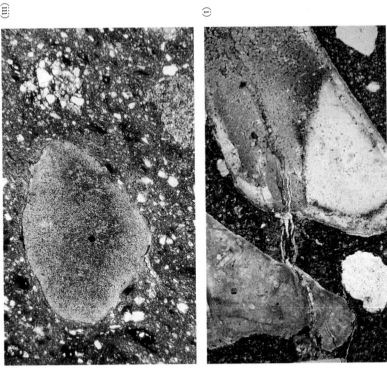

(i) Photomicrograph (taken in plane polarized light) of a concrete in which the alkali-silica reaction has taken place. The field of view is 2.85 × 1.85 mm. The photomicrograph shows a vein, accentuated by the use of a blue-dyed resin for sample impregnation, cutting both flint aggregate fragments and the cement paste matrix.

(ii) Photomicrograph (taken in plane polarized light) of a test brickette prepared from Quaternary glacial till (boulder clay). The field of view is 3.5 × 2.2 mm. Note the variation in grain size shown by the rock and quartz fragments. The blue areas are fractures containing blue-dyed resin used for sample impregnation.

(iii) Photomicrograph (taken in plane polarized light) of a brick manufactured from Carboniferous Etruria Marl. The field of view is 3.5 × 2.2 mm. Note the presence of fine and coarse grained rock fragments set in a matrix of the firing products of the clay component.

(iv) Photomicrograph (taken in plane polarized light) of a brick manufactured from Carboniferous shale. The field of view is 3.5 × 2.2 mm. Note the predominance of coarse grained quartz, derived from crushing of a silty component represented by rock fragments such as that which dominates the right hand side of the image.

Table 4.3 Distribution and age of important sedimentary phosphate deposits shown in Figure 4.7 (from Harben and Bates, 1990; Cook, 1984)

	Location	Age
1	Western USA (Montana, Idaho, Utah, Wyoming, Nevada)	Permian
2	North Carolina, USA	Lower Miocene
3	Florida, USA	Middle Miocene
4	Baja California, Mexico	Upper Jurassic
5	Hidalgo State, Mexico	Jurassic
6	Falcon State, Venezuela	Lower Miocene
7	Sechura Desert, Peru	Upper Miocene
8	Tunisia	Upper Cretaceous–Eocene
9	Algeria	Upper Cretaceous–Eocene
10	Morocco	Upper Cretaceous–Eocene
11	Western Sahara	Upper Cretaceous–Eocene
12	Senegal	Eocene
13	Togo	Eocene
14	Syria	Upper Cretaceous
15	Jordan	Upper Cretaceous
15	Israel	Upper Cretaceous–Eocene
16	Egypt	Upper Cretaceous
17	Kazakhstan	Lower Cambrian
18	Udaipur, India	Precambrian
19	Yunnan, China	Lower Cambrian
20	Lao Kay, Vietnam	Lower Cambrian
21	Vietnam	Lower Cambrian
22	Korea	Lower Cambrian
23	Georgina Basin, Australia	Middle Cambrian
24	Namibia	Upper Cretaceous–Eocene

input, in a warm, arid climate. Anoxic bottom conditions are required, with apatite precipitating as a nodular diagenetic cement within the sediment. If this is reworked, uncemented non-phosphate material is winnowed out, leaving a lag deposit of nodular apatite. There is a strong climatic control on the formation of sedimentary phosphates, being largely restricted to a belt within 40° latitude from the equator (Cook, 1984). There is no restriction in terms of geological age, as phosphate deposits occur in Phanerozoic rocks of all ages, but particularly in the Cambrian, Ordovician, Carboniferous, Permian, Jurassic and late Tertiary–Recent periods (Figure 4.8).

Although very large amounts of phosphate rock are produced each year, over 75% of production is derived from just three sources: the USA, the CIS and Morocco. Because of relatively low domestic consumption, Morocco is the leading exporter of phosphates. Important deposits in North America include those of Florida and North Carolina, which are of Miocene age, and the Permian Phosphoria Formation of Wyoming.

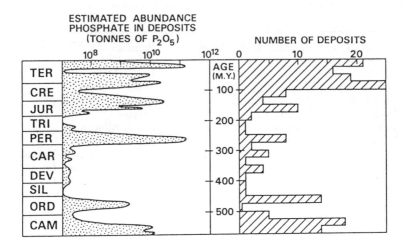

Figure 4.8 Temporal distribution of phosphate deposits (from Cook, 1984).

The Moroccan deposits straddle the Cretaceous–Tertiary boundary (Figure 4.9), and are concentrated into a sequence only 20 m thick but with three beds of particularly high grade (27–32% P_2O_5). The phosphate rock is very friable, and interbedded with cherts, clays and limestones. It is worked both underground (especially where steeply dipping) and in open pits. In some occurrences, the dip of the beds is sufficient to take them beneath the water table, where oxidation has not taken place, preserving their organic matter. These phosphates need to be calcined before use.

Igneous phosphates are in many ways the peculiarities of the igneous rock family. Each deposit seems to have a unique mineral assemblage and mode of occurrence. Examples include those associated with alkaline igneous intrusions, such as the important Khibiny deposits of northern Russia, which form part of a late Carboniferous–early Permian ring complex (Figure 4.10). The apatite occurs as a nepheline–apatite ore within ijolites (nepheline-aegirine-apatite-sphene rocks) with 30%–90% apatite corresponding to 15–45% P_2O_5, with an average grade of 18% P_2O_5. The ring complex is layered, and the apatite ore forms a unit up to 200 m thick, with a strike length of several kilometres, which dips towards the centre of the complex. The apatite is believed to have been concentrated by crystal settling within the magma chamber. Other deposits associated with carbonatites include those of South Africa (Palabora), where the apatite occurs within pyroxenite or with serpentine and magnetite, and those of Finland, where the apatite is found within igneous carbonates.

SULPHUR

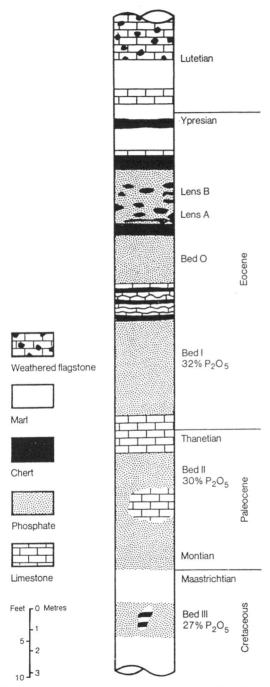

Figure 4.9 Representative vertical succession for the Khouribga phosphate deposit, Morocco (from Harben and Bates, 1990; adapted from Mew, 1980).

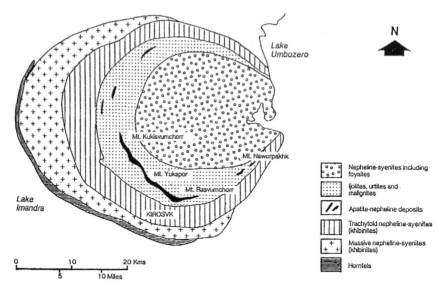

Figure 4.10 Distribution of apatite-bearing lithologies in the Khibiny Complex, Kola Peninsula, Russia (from Harben and Bates, 1990; adapted from Northolt, 1979).

Irrespective of their source, phosphate rocks contain apatite which may be rich in fluorine, and this may have detrimental effects on groundwater quality near any mining operation (or processing plant). Fluorine contents may reach as much as 6% in francolite (Deer *et al.*, 1992). Typically, the fluorine content of groundwater which is regarded as beneficial to health is of the order of 1 mg/L, but amounts in excess of 3 mg/L are hazardous, resulting in teeth and bone disease. Groundwaters with up to 10 mg/L fluoride have been reported from the Gaza Strip, with the possibility that the nearby Negev phosphate deposits are the source of the high fluorine contents.

4.5 SULPHUR

Most sulphur is used in the form of sulphuric acid, which is an essential component of many chemical industrial processes. The biggest single use for sulphuric acid is in the manufacture of phosphate fertilizers, reflecting the scale of the fertilizer industry. It is also used in the petroleum and mining industries, as a catalyst or for acid leaching. Elemental sulphur is traded as a solid or in liquid form, in which case it is transported above its melting point (113°C) in thermally insulated tankers. Solid 'bright sulphur' should contain no more than 0.08% carbon as an impurity, and solid dark sulphur less than 0.25% carbon. Other impurities are also restricted (e.g. arsenic <0.25 ppm, tellurium and selenium <2 ppm), but these restrictions are only relevant to sulphur derived from metal sulphide ores. Canadian

SULPHUR

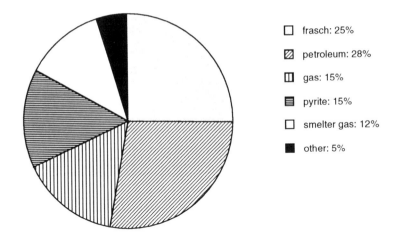

Figure 4.11 World production of sulphur (1990; 53 million tonnes) divided according to origin (data from World Mineral Statistics, 1986–1990; Lofty et al., 1992).

sulphur is available as solid 'slates' produced by casting molten sulphur onto a liquid-cooled conveyer belt, which yields plates some 6 mm thick by 15 cm across.

Although sulphur occurs alone as native sulphur, it is most commonly found as a component of sulphide minerals, such as pyrite (with 53% S), or as a component of petroleum fluids. Within petroleum liquids, sulphur is present in an organic form, but within petroleum gases sulphur may occur as hydrogen sulphide (sour gas). All four types of sulphur occurrence are exploited (Figure 4.11).

4.5.1 Native sulphur

Native sulphur supplies less than 40% of world production, derived mainly from bedded deposits or salt dome cap rock where gypsum or anhydrite have been affected by microbiological activity to produce calcite and native sulphur. Microbes such as *Desulfovibrio desulfricans* consume hydrocarbons as their primary energy source, using sulphur instead of oxygen as a hydrogen acceptor and so producing H_2S. If trapped, this can be oxidized to yield native sulphur, which is commonly found in salt dome cap rocks.

The mining of salt dome cap rocks depends on the Frasch process, in which a well is drilled into the sulphur deposit. Four concentric pipes are placed in the well – the outer to provide a protective casing, followed by one to carry hot water, one to bring sulphur to the surface and the central pipe to carry compressed air. Sulphur melts at 113°C, and is mobilized by the injection of superheated steam and hot water to be brought up to the surface by the influence of the compressed air pressure.

Conventional mining of native sulphur on a significant scale is limited to deposits in Iraq, near Mosul, where Miocene evaporites contain native sulphur, and to deposits in Poland, again within Miocene evaporites. The historically important Sicilian deposits of native sulphur (which have been mined since classical times) are also hosted by evaporite sequences.

4.5.2 Pyrite and metal-sulphide sulphur

Pyrite (FeS_2) is usually described in texts on ore mineralogy, with the implicit assumption that it is mined as metal ore. However, pyrite is an important sulphur ore in its own right (accounting for historical names such as the 'Great Sulphur Vein' in the North Pennine Orefield). It is commonly accompanied by other metal sulphides, in which case sulphur becomes important as a by-product of metal mining and smelting operations. The amounts of sulphur produced from sulphide minerals are very significant indeed: overall, pyrite sulphur accounts for up to 16% of world production, with smelter gas (derived from other metal sulphides) yielding a further 14%.

The major source of pyrite sulphur is from the Iberian Pyrite Belt, which extends from Portugal into southern Spain (Figure 4.12). This region is a major producer of metal ores, from Lower Carboniferous polymetallic sulphide deposits of submarine exhalative volcanic origin. Reserves are enormous, estimated to be some 600 million tonnes of pyrite. The most attractive ore is massive pyrite, with little intergrown contamination by other sulphides, which requires little mineral processing to produce pyrite

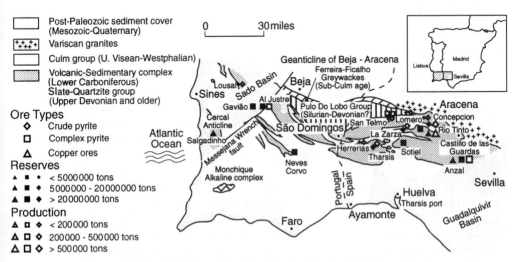

Figure 4.12 Sulphide deposits of the Iberian 'Pyrite Belt' (from Harben and Bates, 1990; adapted from Strauss *et al.*, 1977; 1981).

concentrates. The ores occur in lenses which are worked in large open pits, in steeply dipping deposits approaching 250 m in thickness and in excess of 1 km in horizontal extent.

Sulphur is recovered from sulphide minerals not as the pure element but as the oxide, SO_2, which is produced by oxidative roasting. Sulphur dioxide is the principal feedstock for sulphuric acid production, which involves catalytic oxidation to SO_3 followed by dissolution of the sulphur trioxide in sulphuric acid to produce disulphuric acid ($H_2S_2O_7$). The addition of water hydrolyses the disulphuric acid to produce sulphuric acid:

$$H_2S_2O_7 + H_2O = 2H_2SO_4 \quad (4.6)$$

4.5.3 Recovery from petroleum liquids and natural gas

Petroleum naturally contains sulphur, as a constituent of its organic molecules. During the refining process the oil is 'sweetened' by the removal of sulphur during a hydrogenation step, which produces hydrogen sulphide gas. Hydrogen sulphide is also frequently found as a component of natural gas (sour gas), and in some cases is the dominant gas species. The sour gas resources of the Western Canada Basin are now perhaps more important as a source of sulphur than as a source of methane, and in large part account for the importance of Canada as the world's largest exporter of sulphur. The sour gas contains generally between 1–60% H_2S, with a reported maximum of 87% H_2S in the Panther River area, 110 km north-west of Calgary. Again within the Western Canada Basin, the north Alberta tar sands also represent an important source of sulphur which is recovered during refining of petroleum derived from the sands.

Once generated during refining or produced from natural gas, hydrogen sulphide can be used as a source of sulphur by means of the Claus process. This involves partial burning in a limited air supply to give sulphur dioxide, which is then reacted with unchanged H_2S over an appropriate catalyst:

$$2H_2S + 3O_2 = 2SO_2 + 2H_2O \quad (4.7)$$
$$2H_2S + SO_2 = 3S + 2H_2O \quad (4.8)$$

or (net):

$$2H_2S + O_2 = 2H_2O + 2S \quad (4.9)$$

Note that this results in the formation of elemental sulphur.

4.5.4 Other sources of sulphur

One automatically associates sulphur with volcanic activity, and it may have come as a surprise to find that volcanic sulphur is not a major source of this element. Volcanic sulphur is widespread in its occurrence, but production is limited to small amounts from locally important mines, especially in Japan and Chile.

A further source of sulphur, of limited importance internationally (but historically important in Britain), is from anhydrite, which is treated by the Müller-Kühne process involving roasting ground anhydrite together with clay, sand and coke in a rotary oven at temperatures between 1200°C and 1400°C. The products of this process include cement clinker (see Chapter 7) and sulphur dioxide gas:

$$2CaSO_4 + C = 2CaO + 2SO_2 + CO_2 \quad (4.10)$$

anhydrite   coke   (clinker)

The evolved gases are purified, and the sulphur dioxide so obtained produced into sulphuric acid.

Finally, it is important to consider flue gas desulphurization as a potential source of sulphur. The combustion of sulphur bearing fuels, especially coal, releases sulphur dioxide to the atmosphere and contributes to the acid rain problem by generating dilute sulphuric acid. Flue gas desulphurization offers the potential to harvest this source of sulphuric acid before it enters the atmosphere, and involves scrubbing of flue gases with ground limestone to produce gypsum. The amounts of sulphur involved are very large: in the USA some 600 million tonnes of coal with an average sulphur content of 1.5% are burnt annually, together with 40 million tonnes of oil (average 1.3% sulphur), yielding 15 million tonnes of sulphur equivalent. The products of flue gas desulphurization are unlikely to compete with alternative sources of sulphur, but do offer a potential threat to the gypsum producers.

4.6 ZEOLITES

Naturally occurring and synthetic zeolites are of immense importance to the chemicals industry, principally in their use in production processes, where they find applications as catalysts, molecular sieves, as drying agents and as ion exchangers. Many applications are dominated by artificial zeolites, tailor-made for particular purposes, and natural zeolites may have difficulty in meeting the required specifications. However, new domestic markets include the use of zeolites in washing powders, providing a useful outlet for natural zeolites by virtue of the bulk of material that is required.

The zeolite family includes a very large number of mineral species (45 of which occur naturally, with laumontite, clinoptilolite, heulandite and analcite being most abundant) united by the common property of having an aluminosilicate framework structure based on rings of $(Al,Si)O_4$ tetrahedra. The feldspathoids sodalite and analcite resemble zeolites in their structure and properties, and are included within this category for convenience. Within the framework structure occur cavities which contain

Table 4.4 Variation in zeolite channel dimensions as a function of the number of ring tetrahedra (from Deer, Howie and Zussman, 1992)

Zeolite	Number of tetrahedra within rings	Approximate minimum dimensions of widest channel (Å)
Sodalite	4 and 6	2.2
Analcite	4, 6 and 8	2.8
Phillipsite	4 and 8	4.8 × 2.8
Harmotome	4 and 8	4.4 × 4.2
Levyne	4, 6 and 8	3.2
Erionite	4, 6 and 8	4.2 × 3.6
Chabazite	4, 6 and 8	4.1 × 3.7
Heulandite	4, 5, 6, 8 and 10	7.9 × 3.5
Ferrierite	4, 5, 6, 8 and 10	5.4 × 4.2
Gmelinite	4, 6, 8 and 10	6.4
Mordenite	4, 5, 6, 8 and 12	7 × 6.7
Faujasite	4, 6 and 12	9

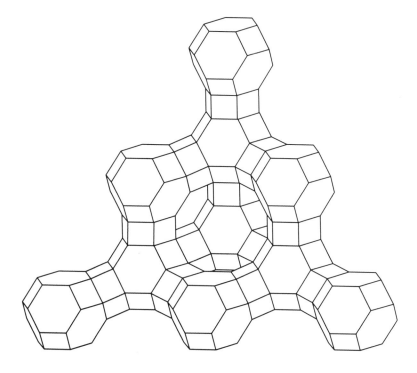

Figure 4.13 The open framework structure of synthetic zeolite A. Each node represents a silicon or aluminium atom, and each line represents a bridging oxygen.

Table 4.5 Compositions of the major zeolite groups and related minerals

Zeolite	Composition
Natrolite	$Na_{16}[Al_6Si_{24}O_{80}].16H_2O$
Thomsonite	$Na_4Ca_8[Al_{20}Si_{20}O_{80}].24H_2O$
Mesolite	$Na_2Ca_2[Al_6Si_9O_{30}].8H_2O$
Laumontite	$Ca_4[Al_8Si_{16}O_{48}].16H_2O$
Analcite	$Na[AlSi_2O_6].H_2O$
Wairakite	$Ca[AlSi_2O_6]_2.2H_2O$
Harmotome	$Ba_2[Al_4Si_{12}O_{32}].12H_2O$
Phillipsite	$K_2(Ca_{0.5},Na)_4[Al_6Si_{10}O_{32}].12H_2O$
Gismondine	$Ca[Al_2Si_2O_8].4H_2O$
Chabazite	$Ca_2[Al_4Si_8O_{24}].12H_2O$
Erionite	$NaK_2MgCa_{1.5}[Al_8Si_{28}O_{72}].28H_2O$
Mordenite	$Na_3KCa_2[Al_8Si_{40}O_{96}].28H_2O$
Ferrierite	$(Na,K)_2Mg[Al_3Si_5O_{36}]OH.9H_2O$
Heulandite	$(Ca,Na_2,K_2)_4[Al_8Si_{28}O_{72}].24H_2O$
Clinoptilolite	$(Na,K)_6[Al_6Si_{30}O_{72}].24H_2O$
Stilbite	$NaCa_2[Al_5Si_{13}O_{36}].14H_2O$

large ions (for charge balance) or water molecules and which interconnect to form channels. The enclosed species are free to move through the channels in the framework structure, permitting ion exchange or reversible dehydration, as long as the size of the potentially mobile species is less than the minimum dimension of the widest channel. This dimension increases with an increase in the number of aluminosilicate tetrahedra which make up the rings of the structure (Table 4.4). A typical zeolite structure is shown in Figure 4.13, which shows the framework which separates the channels through which water or cations can pass.

Natural zeolites are subdivided according to their structures into six groups (Table 4.5 and Figure 4.14):

- Natrolite group: characterized by chain-like structural units and fibrous morphology (Figure 4.14a).
- Laumontite group: characterized by rings of four tetrahedra, which are connected to form chains in which the rings share two single opposite tetrahedra (Figure 4.14b).
- Harmotome group: characterized by rings of four tetrahedra, which connect to form chains in which the rings share opposite pairs of tetrahedra (Figure 4.14c).

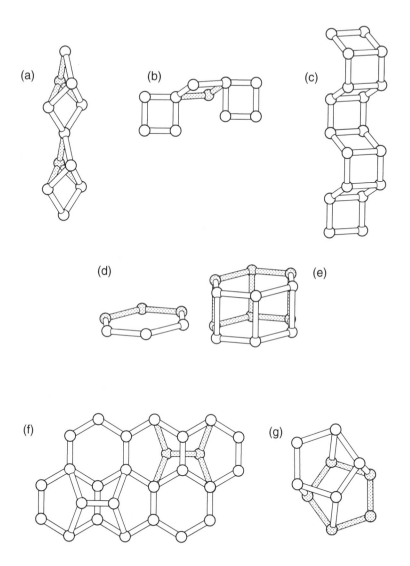

Figure 4.14 Summary of zeolite structural types (based on Deer, Howie and Zussman, 1992). Each node represents an $(SiAl)O_4$ tetratedron. (a) the chain structure typical of the natrolite group, (b) the singly-connected four-ring chain of the laumontite group, (c) the doubly-connected four-ring chain structure of the harmotome group, (d) the single six-ring structure characteristic of erionite and (e) the double six-ring structure of chabazite, (f) the six-ring sheet structure, with 'handles', of the mordenite group and (g) the five-membered ring structure of the heulandite group.

- Chabazite group: contains single or double rings composed of six tetrahedra oriented perpendicular to a three- or six-fold axis of symmetry to give an open columnar structure (Figure 4.14d, e).
- Mordenite group: contains six-membered rings arranged in a sheet structure, with additional pairs of tetrahedra located on either side of the sheet and acting as a link between sheets to give a three dimensional structure (Figure 4.14f).
- Heulandite group: characterized by a structure based on three- and five-membered rings (Figure 4.14g).

Chemically, although their formulae appear to be forbidding, all zeolites are characterized by constant molecular ratios for the alkalis and alkaline earths to alumina ((Ca, Sr, Ba, Na_2, K_2)$O:Al_2O_3 = 1$) and Al + Si to oxygen (($Al + Si$)$:O = 1:2$). In the synthesis of zeolites, the proportion of Si to Al can be varied to give very silica-rich species which do not occur in nature.

4.6.1 Occurrence of zeolites

Zeolites are commonly found in association with volcanic rocks, often forming well-developed crystals within cavities. They also occur less spectacularly but in much greater quantity as components of volcanic ashes interbedded within saline alkaline lake deposits, in marine tuffs, and are produced during metamorphism. Zeolites are widely produced as a consequence of low temperature alteration of sediments rich in volcanic material, and individual zeolite minerals are characteristic of particular conditions, especially temperature. Phillipsite forms at the lowest temperatures, precipitating at 4°C at the sediment–water interface in deep sea sediments, and continuing to grow within the sediment. It is restricted in general to Miocene and younger sediments. Clinoptilolite, mordenite and analcite are formed at successively higher temperatures.

Commercial occurrences of zeolites are dominated in significance by those of Japan and the USA, with important production from Hungary, Bulgaria, west Germany, Cuba, New Zealand, Iceland, Italy and South Africa. The Japanese deposits consist of thick clinoptilolite-bearing tuffs of Miocene age, composed of marine rhyolitic and dacitic ash which have undergone very low-grade hydrous metamorphism to produce the zeolite. In the USA, the most important zeolite deposits occur in Arizona, within Pliocene–Holocene tuffs deposited in an intermontane saline lake environment. Chabazite forms up to 80% of certain beds, occurring with clinoptilolite and erionite.

Fired products: the need for high temperature processing   5

Many industrial products derived from mineral raw materials are made by heating the raw materials, alone or in appropriate proportions, so that they are transformed into a new material which has the physical and/or chemical properties that are required by the manufacturer. Essentially, two types of process are involved:

1. heating to produce a chemical change, yielding a material which is designed to be very reactive chemically when in use, such as plaster or cement.
2. heating to produce both a chemical and a physical change, and yielding a material which is designed to be relatively unreactive chemically under the conditions of use – such as a refractory brick, glass or building bricks.

The firing of industrial minerals is necessarily an energy-intensive process, covering a range in temperature from as little as 150°C (in plaster manufacture) to 1600°C (in glass manufacture). Large amounts of material are usually involved, in batch or continuous processes (Table 5.1).

Table 5.1 Production statistics (1989) for materials which require firing (source: CEC Panorama of EC Industry 1991–1992)

Product	UK	EU*	USA	Japan
Glass (million tonnes)	3	21	15	13
Cement (million tonnes)	15	174	70	82
Bricks (cubic metres)	6.5×10^6	31×10^6	–	–

Note that comparability of brick statistics is complicated by difficulties in relating variation in brick size, volume and density between countries; * the former EEC; includes data for the UK.

In order to save energy and to reduce waste by poor formulation of the batches of raw materials that are to be fired, it is essential that the process of firing is understood as fully as possible, so that the effects can be predicted for a given set of circumstances. In general, high temperature processing of bulk materials is carried out at atmospheric pressure, but a number of materials are produced at elevated pressure, under autoclave conditions in giant pressure cookers. High pressure production is usually carried out on a batch basis, and is not generally amenable to continuous processes, largely because of the technical difficulties involved in manipulating pressure gradients.

5.1 PREDICTION OF THE EFFECTS OF FIRING

The behaviour of minerals at high temperatures and at high pressures has long been of interest to geologists and materials scientists. The two groups have worked very closely together during throughout the twentieth century to develop a quantitative understanding of mineral phase relationships under either manufacturing or geological conditions. Eminent scientists such as N.L. Bowen contributed both to geology and materials science, publishing in the literature for either field with equal authority; Bowen's book, 'The Evolution of the Igneous Rocks', originally published in 1928 and reprinted in 1956, is still a seminal work well worth reading (Bowen, 1956).

The information which is available to help predict the behaviour of mineral raw materials at high temperatures and pressures includes information bearing on reactions which take place during heating and details of the conditions at which melting occurs. The standard way of presenting this information is as a **phase diagram**, which can be drawn to show mineral stability fields as a function of selected variables such as temperature, pressure and composition. Phase diagrams for several important chemical and mineralogical **systems** (which are defined chemically) have been accumulated since the 1920s, and are now widely available for most systems of interest. In principle, they are very simple to determine by means of carefully designed experiments. The experimentalist takes the charge, a chemical or mineral mixture of the required composition, subjects it to the pressure and temperature conditions of interest for a period of time which is sufficient for reaction to go to completion, quenches the charge and analyses the products. This is done for a number of selected compositions and conditions, so that the results can be plotted in diagrammatic form, and then the plotted points can be used to produce a working phase diagram. In practice, it isn't as simple as it sounds! First, care has to be taken in the preparation of the charge – some starting materials are more reactive than others, so a synthetic gel (an amorphous precipitate of the required chemical composition; Hamilton and Henderson, 1968) may be preferable to a mechanical mixture of ground mineral fragments. Secondly, the charge has to be held in a furnace at the required temperature, without reacting with the container – it may be necessary to use gold or platinum crucibles or capsules rather than a ceramic crucible. Third, if high pressures are required, the charge and its container have to be immersed within a pressure vessel, surrounded by a pressure medium such as water, hydraulic oil, gas or a deformable solid, without rupture of the sample container or contamination of the charge. For very high pressures, the equipment is usually substantial in its construction, and the sample volume might be very small (a few cubic millimetres) to allow the required pressure to be reached using apparatus which can conveniently fit in a laboratory. Some equipment takes a long time to heat up and then to cool down at the end of an experiment, and so quenching the charge to preserve the assemblage which is present at high temperature and pressure might be a problem. At the end of the experiment, the analysis of the products might need sophisticated analytical equipment. Even then, the interpretation of the results has to be carried out with considerable care, to ensure that the the products have achieved chemical equilibrium and that no experimental artefacts have affected the results.

Because of all these difficulties, experimental mineralogy is a challenging subject which attracts rather few devotees nowadays. Fortunately for the industrial mineralogist much critical work has now been completed.

5.2 THE INTERPRETATION OF MINERALOGICAL PHASE DIAGRAMS

Mineralogical phase diagrams are very powerful tools which can be used to predict the firing behaviour of mineral raw materials, or conversely the behaviour of natural processes of metamorphic or igneous origin. Most valuable are those which simply relate composition and temperature. The validity of a phase diagram can be assessed using the phase rule:

$$p + f = c + 2 \tag{5.1}$$

where p = the number of mineral phases
where f = the number of degrees of freedom
where c = the number of chemical components; the simplest chemical species which, when mixed in appropriate proportions, can describe any composition in the system.

The number of degrees of freedom refers to the freedom to change the **intensive** variables which describe the system, such as pressure, temperature and composition. However, composition is not a variable in its own right; it is described by other variables such as the proportion of a given component. Because of the need for proportions to add up to one (or percentages to add up to 100), all but one of the compositional variables must be stated, as the last is constrained by the total. Thus for a three-component system the minimum requirement to describe any composition is to define the proportions of two components. Intensive variables are those which are independent of the size of the system, whereas **extensive** variables depend on the size of the system. Knowledge of extensive variables, such as the weight of material present altogether, is not required.

5.2.1 Single-component systems

Figure 5.1 shows relationships for phases stable in the sytem SiO_2. It illustrates that there are a number of silica polymorphs: high (or α) and low (or β) quartz, tridymite, cristobalite and liquid. This is a one-component system, in that all of the phases are composed simply of SiO_2, and there is

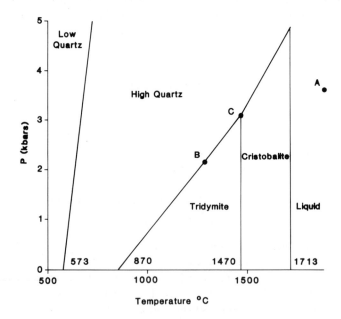

Figure 5.1 Phase diagram for the one-component system SiO_2, showing the stability fields of the silica minerals as a function of changes in pressure and temperature (see text for discussion).

no need to state the composition. Application of the phase rule can be illustrated as follows:

At point A, *within* the liquid field, there is one phase (liquid) so p equals 1, one chemical component (SiO_2) so c equals 1 and thus there are two degrees of freedom as $p + f = c + 2$. In other words, both pressure and temperature are free to vary independently without affecting the number of phases present. The same situation holds for any point within the other stability fields where one phase is present.

On a **field boundary**, at point B, p now equals 2 (high quartz and tridymite coexist in equilibrium), c still equals 1, and so f has to equal 1. There is only one degree of freedom, and so if temperature varies then pressure must also vary in order to maintain equilibrium between the two phases and the plotted position must remain on the field boundary.

At a **triple point** (point C), p now equals 3 (high quartz, tridymite and cristobalite all coexist in equilibrium). The number of components, c, is unchanged at 1, and so f has to equal 0. Pressure and temperature are both fixed; neither can be varied without losing the equilibrium between the three phases.

The system SiO_2 is of practical importance in the manufacture of silica bricks for refractory purposes, and we will refer to it in a later chapter. Other single component systems of geological importance include the aluminium silicate diagram (phase relationships for the system Al_2SiO_5, which can be considered for convenience as a single component system), and (of course) the system H_2O, which illustrates phase relationships for ice, water and steam. The use of the phase rule, as illustrated above for the system SiO_2, can be applied equally to these and other single component systems.

5.2.2 Two-component systems

In a two-component system, composition becomes a variable that has to be considered in addition to temperature and pressure, and is expressed as the proportion of one component (percentage or fraction) which fixes the proportion of the other component. It is inconvenient to plot three dimensional figures on a two dimensional piece of paper, and so it is general for two-component systems to be described at a constant pressure (which should always be stipulated). Phase diagrams for two-component systems are generally constructed so that the phase relationships can be described graphically as a function of temperature and composition. Again, the phase rule helps to understand the way in which the diagram can be interpreted, but if pressure (or another independent variable) is held constant one degree of freedom is removed and the phase rule becomes:

$$p + f = c + 1 \tag{5.2}$$

Figure 5.2 shows phase relationships in the two-component system SiO_2–MgO at atmospheric pressure. It illustrates that, in addition to liquid, there are a number of mineral phases within this system:

- the pure **end-members** periclase (MgO) and a silica **polymorph** (low quartz, high quartz, tridymite or cristobalite, depending on temperature)
- the **intermediate compounds** forsterite (Mg_2SiO_4; an olivine) and enstatite ($MgSiO_3$; a pyroxene)

The phases which occur at a given set of temperature and compositional conditions can be deduced from the diagram, again applying the phase rule: At point A, within the liquid field, $p = 1$, $c = 2$ and there are two degrees of freedom, as pressure is constrained. In other words, either temperature or composition can vary; if one varies, the other can be varied independently or not at all, with no change in the resulting phase assemblage. However, in a closed system, the bulk composition is fixed, and one of the two degrees of freedom is lost. Consequently, variation is restricted to changes in temperature. This is of course highly appropriate, reflecting the heating or cooling of the material of interest. Again, the same situation holds for any point within the other phase stability fields where composition is fixed.

On the **liquidus** curve (the curve which represents the upper limit for the existence of crystalline phases) at point B, p now equals 2 (periclase and liquid coexist in equilibrium). c still equals 2, and so f must equal 1. There is only one degree of freedom, and so if temperature varies then liquid composition must also vary, along the field boundary, in order to maintain equilibrium between the two phases, as the composition of the mineral is fixed. Liquid composition varies by precipitating or dissolving the crystalline phase, in this case periclase, as temperature falls or rises.

At point C, which is a **eutectic point**, p now equals 3 (periclase, forsterite and liquid all coexist in equilibrium), c is unchanged at 2, and so f must equal 0. Like pressure, temperature is now fixed and cannot be varied without losing the equilibrium between the three phases. If cooling is continued, it will only take place once the liquid has been consumed completely by freezing. The temperature which corresponds to the eutectic point represents the temperature limit below which no liquid can occur for compositions intermediate between those of periclase (MgO) and forsterite (Mg_2SiO_4), or conversely the upper temperature limit for fields in this compositional range composed only of solid phases. For compositions richer in silica (between Mg_2SiO_4 and SiO_2) there is another eutectic at a lower temperature (point D). The horizontal line drawn through a eutectic is called the **solidus**. This system can be subdivided into two sub-systems:

INTERPRETATION OF MINERALOGICAL PHASE DIAGRAMS

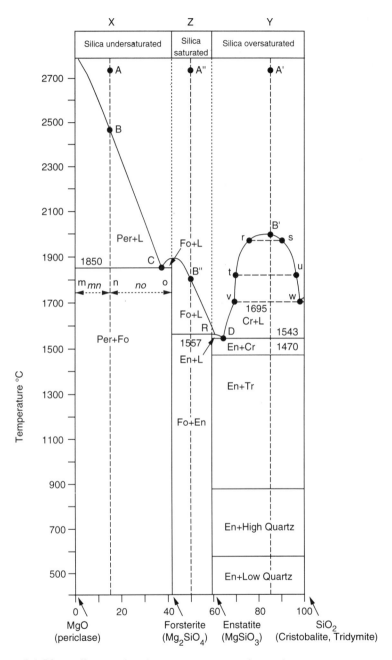

Figure 5.2 Phase diagram for the two-component (binary) system SiO_2–MgO, at atmospheric pressure, showing the stability fields of minerals containing Mg, Si and O as a function of bulk composition (expressed in weight %) and temperature (see text for discussion).

periclase–forsterite and forsterite–silica, each of which has a eutectic, at temperatures of 1850°C and 1543°C respectively. Below the solidus, horizontal lines indicate the temperatures at which polymorphic transitions (or reactions between solid minerals) take place. In this example, the transitions involve the silica polymorphs, and correspond to the temperatures shown in Figure 5.1 for a pressure of 1 bar.

A two component diagram can be used to illustrate the mineral reactions which take place during cooling or heating of a given composition. In Figure 5.2, the cooling of a composition X (15% SiO_2, 85% MgO) can be predicted as follows:

Liquid alone is present (point A) until the cooling path intercepts the liquidus for periclase at point B. We are now on a field boundary, and so f equals 1; we can only proceed to lower temperatures if a composition changes. Periclase has a fixed composition, and so continued cooling must involve a change in the liquid composition. The following happens: the liquid becomes depleted in MgO as it is removed by the precipitation of periclase, and the trajectory in the phase diagram of the liquid composition is along the liquidus boundary towards the eutectic C. At all times the bulk composition X is preserved, but the proportions of liquid and crystals changes. At point C cooling cannot continue until all of the liquid is consumed, and this requires that forsterite precipitates, in order to remove the remaining silica within the melt as well as the stoichiometric proportion of MgO (2 moles of MgO for every mole of SiO_2). The trajectory now continues within the field for periclase + forsterite, and involves a mixture of the two minerals. The proportion of the two phases can be determined using the *lever rule*. If a horizontal line is drawn through the cooling trajectory, to form a lever, the proportions of the phases which are present can be measured using the relative lengths of the lever between the composition of interest (the fulcrum) and the end member or intermediate compositions which are first intercepted moving in either direction along the lever away from the fulcrum. Referring to Figure 5.2, this can be expressed as:

$$\% \text{ periclase} = no \div (mn + no) \, (\%) \quad (5.3)$$

and

$$\% \text{ forsterite} = mn \div (mn + no) \, (\%) \quad (5.4)$$

Note that the proportion of a constituent phase is represented by the opposite length of the lever – more obvious intuitively when compositions close to either pure phase are considered.

The phase diagram for the system MgO–SiO_2 shows many additional features, most of which will be discussed below when examining its application to refractory products. However, it is important to note here that there is a two liquid field shown for silica-rich compositions (on the

INTERPRETATION OF MINERALOGICAL PHASE DIAGRAMS

right hand side of the diagram, defined by the **solvus** forming an arc linking points v and w). This is a region where **liquid immiscibility** occurs, with a silica-rich liquid and a magnesia-rich liquid coexisting at equilibrium. It is directly analogous to the immiscibility of oil and water in mayonnaise, and when experimental charges are prepared in this region and quenched the resulting glass is opalescent and milky white. Liquid immiscibility of this type is exploited to produce opaque or opalescent glasses, as will be seen later, but it also has to be avoided in other circumstances. The cooling trajectory of a composition Y (15% MgO, 85% SiO_2) which intercepts the two-liquid field follows the dictates of the phase rule. Liquid alone is present until the two-liquid field boundary is intercepted at B'. The liquid then subdivides into two liquid phases (r + s; t + u etc.), and the trajectory splits to follow the two-liquid field boundary down to the liquidus, where cristobalite starts to precipitate from each liquid at v and w. If chemical equilibrium is maintained, cristobalite will start to crystallize from each liquid phase at exactly the same temperature, but in different proportions. Once the silica-rich liquid, of composition w, has been consumed by the precipitation of cristobalite at 1695°C the cooling trajectory will proceed down the liquidus curve to the eutectic D. Further cooling will only take place when the remaining liquid is completely consumed, and the bulk composition will be represented by a constant proportion of enstatite and the silica polymorph which is appropriate for the temperature.

It is also important to note that the system MgO–SiO_2 contains a reaction point, R. This is where the horizontal boundary which separates a field for forsterite + liquid from a field for enstatite + liquid intersects the liquidus curve. When a composition such as composition Z cools, it is first a single liquid (as at point A''), and when it meets the liquidus curve at B'' forsterite crystallizes. On further cooling the liquid composition follows the liquidus curve as more forsterite precipitates, until it meets the reaction point (or **peritectic**) R. Here forsterite reacts with the liquid, which is completely consumed by the reaction, to form enstatite, resulting in a solid assemblage of forsterite + enstatite. For compositions richer in silica than the enstatite composition, but still intersecting the liquidus to the left of the reaction point, some liquid remains (although forsterite is completely consumed by reaction) and continued crystallization leads to point D. Here the liquid is finally consumed as cristobalite crystallizes, resulting in a solid assemblage of enstatite + cristobalite.

5.2.3 Three-component systems

In three-component systems the problems of representing phase relationships on a sheet of paper become even greater than for two-component systems, as there is greater scope for compositional variation. It is again usual to consider phase relations at constant pressure, and it is convenient

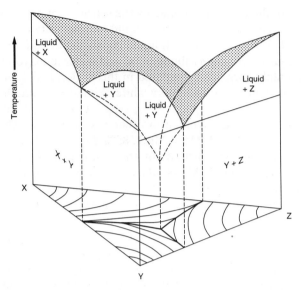

Figure 5.3 Sketch of the three-dimensional temperature–composition model of a hypothetical ternary system X–Y–Z. Each of the sides is a binary system, and the basal triangle carries temperature contours and field boundaries projected from the top surface.

to draw up triangular compositional diagrams in which each of the three components is represented by the corner of the triangle, and which contain temperature information as contours. The representation of phase relationships for a three component system as a triangular diagram is analogous to considering the diagram as a map, in which the contours indicate temperature, and the boundaries between the mineral stability fields can be thought of as valleys, with a downhill direction indicated by an arrow drawn on the boundary (or deduced from the contour pattern). Three-component diagrams can be related to the three two-component systems which constitute their edges by contemplating a three-dimensional solid model in which the three sides each represent a binary system and the contoured top surface is projected onto the basal triangle to give the ternary liquidus diagram (Figure 5.3).

A three-component diagram gives information relating to both the liquidus phase, the first mineral to precipitate on cooling of a liquid, and the solid phase assemblage. This information is usually given on a single diagram. However, in consideration of the system SiO_2–MgO–CaO as an example of a three-component system, liquidus and solidus phase relationships are drawn separately in Figures 5.4 and 5.5, respectively. Figure 5.4 shows the liquidus field ***boundary curves*** for the system SiO_2–MgO–CaO, and the liquidus surfaces which have been contoured according to tempera-

INTERPRETATION OF MINERALOGICAL PHASE DIAGRAMS

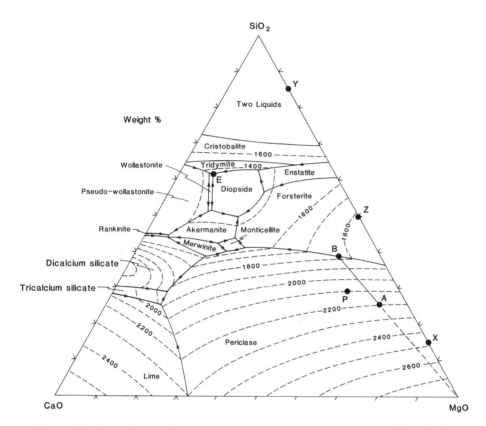

Figure 5.4 Phase diagram for the three-component (ternary) system SiO_2–MgO–CaO, at atmospheric pressure, showing the liquidus fields for minerals (and artifical compounds) containing Mg, Ca, Si and O (Levin, Robbins and McMurdie, 1964). The contours indicate the temperature (°C) of the liquidus surface, and arrows show the 'downhill' direction followed by the liquid composition as it cools (see text for detailed discussion). This diagram is plotted with compositions expressed as weight %.

ture. The sides (*joins*) of the diagram are represented by the binary systems SiO_2–MgO (shown in Figure 5.2), SiO_2–CaO and CaO–MgO. Thus compositions X, Y and Z on the SiO_2–MgO join are identical to those discussed with reference to Figure 5.2, and the behaviour of these compositions can be used to help interpret the behaviour of similar compositions which contain a proportion of lime (CaO). Figure 5.5 plots the compositions of the many mineral phases which are known to be made up of the three components MgO, CaO and SiO_2 in various proportions: their compositions are given in Table 5.2. Tie lines have been drawn between mineral compositions which can coexist, and the extent of solid solutions is shown by thickening these lines for the solid phase composi-

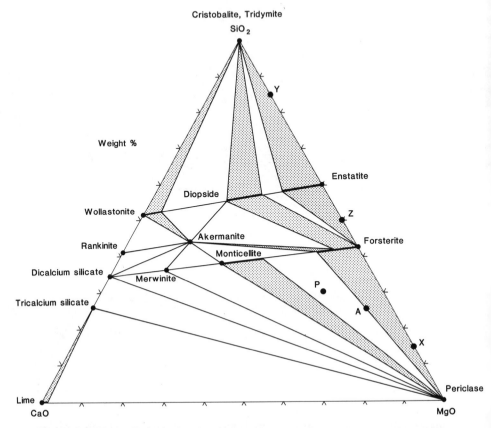

Figure 5.5 Phase diagram for the three-component (ternary) system SiO$_2$–MgO–CaO, at atmospheric pressure, showing the solid phase assemblages for minerals (and artificial compounds) containing Mg, Ca, Si and O (Levin, Robbins and McMurdie, 1964). Details of the compositions of the solid phases are given in Table 5.2, and the diagram is discussed more fully in the text. This diagram is plotted with compositions expressed as weight %.

tional ranges that show solid solution, and by shading the areas in which solid solution affects coexisting solids. The tie lines define compatibility triangles which indicate which solid phases can coexist at equilibrium. Figure 5.6 shows how the compatibility triangles can be read: if a composition plots on a line (e.g. point s) it consists of the two phases at each end of the line (A + B). If it lies within a triangle (e.g. point t) it consists of the three phases forming the corners of the triangle (A + B + D). If it lies on a point (e.g. point u) it consists of a single phase (D).

Referring to Figures 5.4 and 5.5, the compositions X, Y and Z which were used to illustrate cooling behaviour in the binary system SiO$_2$–MgO are shown, plotting on the SiO$_2$–MgO join. The phase boundaries shown in

INTERPRETATION OF MINERALOGICAL PHASE DIAGRAMS

Table 5.2 Crystalline phases in the system $CaO-MgO-SiO_2$

Name	Formula
Cristobalite	SiO_2
Tridymite	SiO_2
Pseudo-wollastonite	$\alpha\text{-}CaSiO_3$
Wollastonite	$\beta\text{-}(Ca,Mg)SiO_3$
Rankinite	$Ca_3Si_2O_7$
Lime*	CaO
Periclase	MgO
Forsterite	$(Mg,Ca)SiO_4$
Protoenstatite	$(Mg,Ca)SiO_3$
Diopside	$(Ca,Mg)_2Si_2O_6$
Akermanite	$Ca_2MgSi_2O_7$
Merwinite	$Ca_3MgSi_2O_8$
Monticellite	$(Ca,Mg)_2SiO_4$
Dicalcium silicate*	Ca_2SiO_4
Tricalcium silicate*	Ca_3SiO_5

* not known to occur naturally.

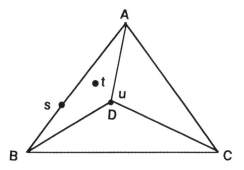

Figure 5.6 Diagram of the use of compatibility diagrams to identify the mineral phases which are stable for a given composition (see text for details).

the ternary diagram are consistent with those shown in the binary diagram (Figure 5.2). For compositions which plot within the diagram (i.e. with the addition of CaO) crystallization involves additional mineral phases or solid solutions in which Ca occurs within forsterite or enstatite. Figure 5.5 shows that both minerals can accommodate significant amounts of calcium within their structures, without losing their status as single phases. Using Figure 5.5, below the solidus composition A consists of a mixture of periclase and forsterite, in which the forsterite is as rich in Ca as it can be. The cooling history of composition A can now be explained as follows: at a temperature above the liquidus surface (i.e. above 2200°C; Figure 5.4) it is completely molten. When applying the phase rule, four variables are needed to describe the point of interest: pressure (which is constant), temperature, the proportion of one component and the proportion of a second component. Using $p + f = c + 1$ (neglecting pressure), the number of degrees of freedom within the single phase liquid field is three; temperature and the two compositional parameters can be varied independently, within limits, without changing the phase relations. However, for a closed system the composition is fixed, losing two of the three degrees of freedom, and so only temperature can vary; A can be heated or cooled without losing its status as a single phase. Once the composition cools to intersect the liquidus, a single mineral phase crystallizes whose identity is given by the label on the liquidus field, in this case periclase. Now there are two phases, so $p = 2$, and a further degree of freedom has been lost ($f = 2$). On further cooling temperature changes, and the composition of the liquid must change. Being constrained by the precipitation of periclase, it follows a straight line trajectory leading away from the periclase composition, the MgO apex, until it intersects the boundary curve which separates the forsterite liquidus field from the periclase liquidus field at point B. In Figure 5.4, this field boundary carries an arrow which shows the 'downhill' direction, confirmed by inspection of the temperature contours. Forsterite begins to crystallize as well as periclase, and further cooling proceeds along the field boundary. An additional phase is present (so $p = 3$), and so there is just one degree of freedom. This means that, on further cooling, the liquid composition changes (but lies on the boundary curve), as periclase and forsterite both crystallize. As it crystallizes, the forsterite becomes increasingly enriched in Ca, until it reaches a composition lying on a straight line drawn through the forsterite composition, A, and the periclase apex of the triangle (Figure 5.5). The liquid is now completely consumed, and the phase assemblage is solid.

If we consider a composition which contains more CaO, such as P (Figure 5.5), the crystallization behaviour will be similar but will continue with the precipitation of monticellite once forsterite is saturated with respect to Ca. When solid, composition P will consist of a mixture of monticellite, forsterite and periclase.

Figure 5.4 shows the temperature contours for the liquidus in the system SiO_2–CaO–MgO, and it is important to note the lowest liquidus temperature, in this case 1320°C (point E), which represents a *ternary eutectic* for the system. No melt can exist for compositions in this system at lower temperatures than this. There are a number of eutectics within this system; they can be identified as points where three downhill arrows converge. The solidus generally is not shown on a ternary diagram; it represents the surface below which no melt can exist, or the upper surface for fields which only contain solids. For eutectic compositions there is no melting interval,

INTERPRETATION OF MINERALOGICAL PHASE DIAGRAMS

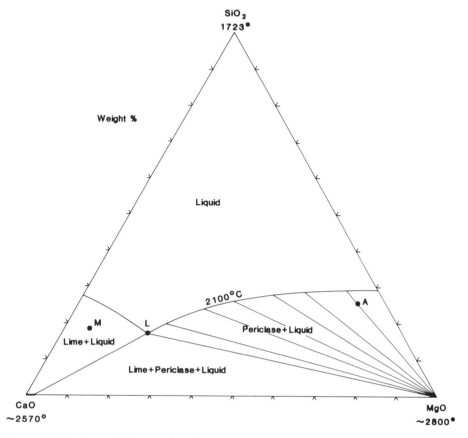

Figure 5.7 Isothermal diagram for the three-component (ternary) system SiO_2–MgO–CaO, at atmospheric pressure, showing the phase assemblages stable at 2100°C (see text for detailed discussion). This diagram is plotted with compositions expressed as weight %.

and eutectic melts crystallize immediately to give the solid phase assemblage indicated by the compatibility triangles shown in Figure 5.5.

It is often the case that the process of interest to an industrial mineralogist involves partial melting, in which case it is necessary to construct an *isothermal section* for the system concerned at a particular temperature. Preparation of an isothermal section is illustrated using Figure 5.7, again for the system SiO_2–MgO–CaO, at a temperature of 2100°C. This is a hypothetical case, in that very few processes (or experiments) are undertaken at such high temperatures, but it is within a simple part of the diagram where the principle can be illustrated easily. Point L represents the position where the periclase–lime field boundary intersects the 2100°C contour. This contour is now drawn as a boundary, separating fields whose liquidus temperatures are below 2100°C from those with higher liquidus temperatures. The fields which are now important are those defined by the triangle periclase–lime–L and its neighbours, both of which have the temperature contour as one boundary. Within the triangle periclase–lime–L all bulk compositions will be made up of those three phases, in different proportions, and the composition of the liquid, L is given by the position of that point. In the neighbouring fields, bulk compositions consist of either periclase (as for point A) or lime (as for point M) together with the liquid whose composition is defined by the intersection of a straight line drawn through either A or M to lime or periclase and the 2100°C temperature contour. In this example, it would be convenient to neglect the limited solid solution of MgO in CaO, as the exact composition of the mineral phase within the solid solution would have to be determined experimentally for an exact estimate of the liquid composition.

There is very much more that can be done to interpret and use phase diagrams such as those presented here, and other books (such as Ehlers (1972) and West (1982)) should be consulted for more detailed study. However, it must be remembered that phase diagrams are generally determined by experiment, and so although they appear to be authoritative they are subject to errors arising from experimental weaknesses. In the three-component diagram used here, for example, it is not possible to undertake experiments at high enough temperatures to allow precise determination of the liquidus for the lime and periclase fields, and so these contours represent estimates; in some cases they can be calculated theoretically. Compilations of phase diagrams are however published by the American Ceramic Society and American Foundry Institute (Levin *et al.*, 1964; Muan and Osborn, 1965), or they can be calculated as required using computer software such as MTDATA (produced by the National Physical Laboratory).

5.3 PRACTICAL APPLICATION OF PHASE DIAGRAMS

The use of phase diagrams generally involves the need to understand the behaviour at high temperatures of rock and mineral raw materials which may have been blended and for which chemical compositional data are known. The first problem to be solved is how to express the composition of the material in a relevant phase diagram, especially as the raw material composition might be complex, with several oxides present in significant quantities. It is commonly the case that more than one three-component diagram will need to be considered to predict firing behaviour, and although the presence of some minor constitents can have a very great effect on phase relationships they can at first be neglected. Take, for

PRACTICAL APPLICATION OF PHASE DIAGRAMS   113

Table 5.3 Compositional data for a hypothetical brick clay (wt %)

Component	1	2	3	4	5	6
SiO_2	63.3	67.3	75.9	75.4	76.4	75.6
Al_2O_3	16.8	18.0	20.3	20.2	20.4	20.2
FeO	3.7	3.9	–	4.4	–	–
CaO	2.6	2.8	–	–	3.2	–
MgO	3.2	3.4	3.8	–	–	–
K_2O	3.5	3.7	–	–	–	4.2
TiO_2	0.7	0.7	–	–	–	–
MnO	0.1	0.1	–	–	–	–
P_2O_5	0.1	0.1	–	–	–	–
LOI	6.0	–	–	–	–	–
Total	100.0	100.0	100.0	100.0	100.0	100.0

N.B. Iron is expressed here as FeO rather than Fe_2O_3.

example, a brick clay with the composition given in Table 5.3. The compositional data are expressed as oxide weight percentages, and include a figure for loss on ignition (LOI; column 1), which represents the amount of water and organic matter that can be lost from the sample when it is ignited at 1000°C (a standard analytical procedure). Of course, this material would also be lost when the clay is fired to produce bricks, and so it can be ignored; column 2 contains the remaining analytical data recalculated to 100%. Silica and alumina dominate, with lesser amounts of iron, calcium, magnesium and potassium oxides. Titanium, manganese and phosphorus oxides together amount to less than 1%, and can be neglected. However, the remaining six components should be considered in any application of phase diagrams to predict how the clay will behave on firing. A further simplification can be made if it is assumed that iron remains reduced and can be expressed as FeO. It is now appropriate to consider plotting the compositional data in each of the four three-component systems SiO_2–Al_2O_3–MgO, SiO_2–Al_2O_3–FeO, SiO_2–Al_2O_3–CaO and SiO_2–Al_2O_3–K_2O. We will use the first one of these for illustrative purposes, but in each case the values for each of the oxides first have to be normalized to 100% (Table 5.3; columns 3–6). Note that the effect of normalizing yields only slightly different figures in each case, reflecting the coincidental similarity of values for each of the components FeO, MgO, K_2O and to a lesser extent CaO.

The composition of the brick clay is plotted in the system SiO_2–Al_2O_3–MgO in Figure 5.8, using the graduations on the edge of the triangle, which are counted in each case towards the apex of interest from the opposite edge, ruling a line parallel to the opposite edge. The intersection of the three lines indicates the position of the point to be plotted. Of course, because the three numbers add up to 100, only two lines need be plotted as all three must share the same intersection – if they do not there is an error

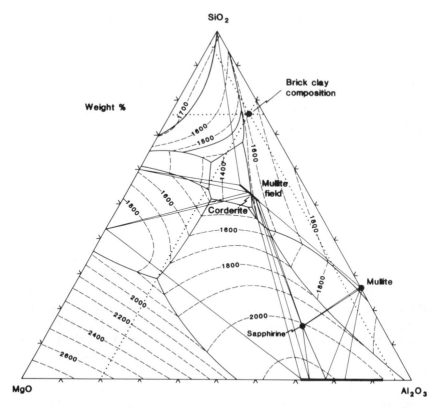

Figure 5.8 Phase diagram for the three-component (ternary) system SiO_2–MgO–Al_2O_3, at atmospheric pressure (Levin, Robbins and McMurdie, 1964), showing the position of a hypothetical brick clay composition. A fully annotated version of this diagram is given in Appendix A. This diagram is plotted with compositions expressed as weight %.

in plotting or recalculating the normalized values. It is possible to determine the composition of a point of interest by reversing this procedure. The lever rule can be applied both to binary mixtures (i.e. those which lie on a straight line linking two end-member compositions), and to three-component mixtures. In the case of a three-component mixture (Figure 5.9) the lever rule involves determination of the relative lengths of lines drawn through the point of interest from the corners of the triangle. The proportion of X in composition c is given by the ratio cx/Xx, of Y by the ratio cy/Yy, and of Z by the ratio cz/Zz. In many cases it is easiest to work out the proportions of two of the three components by projecting from one corner through the point of interest to the opposite edge. For example, a line can be drawn from Z through c to z, and then using the lever rule the relative proportions of X and Y in z are estimated (i.e. the

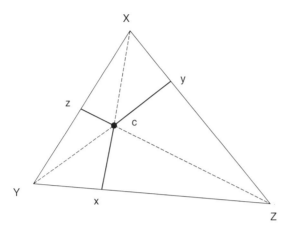

Figure 5.9 Application of the lever rule to triangular diagrams (see text for discussion).

proportion of x is given by the ratio zy/xz). The lever rule is then used again to calculate the proportion of this X + Y mixture in c, given by the ratio Zc/Zz, and the amounts of X and Y can then be calculated individually.

An initial interpretation of the plot shown in Figure 5.8 is that the brick clay lies within the mullite liquidus field, and that the composition can be expressed as a mixture of mullite, a silica mineral (cristobalite, tridymite or quartz, depending on the temperature) and cordierite. The predicted temperature of the liquidus is approximately 1620°C. This diagram, together with others relevant to brick manufacture, will be discussed again in Chapter 8.

Mention has already been made of the loss of volatiles when a material is ignited. Of particular importance is the loss of carbon dioxide from carbonate rocks and minerals, which are raw materials of major importance. They are very convenient as raw materials; it is possible to find carbonate rocks of high purity, and on ignition they lose carbon dioxide leaving behind an oxide component which might not occur freely in nature, as in Reaction 5.5:

$$CaCO_3 = CaO + CO_2 \qquad (5.5)$$
$$\text{calcite} \quad \text{lime} \quad \text{carbon dioxide}$$

The oxides which remain are highly reactive, and combine with other components of blended raw materials to either melt or produce a mineral product of the required composition.

When applying phase diagrams to carbonate rocks and minerals, it is important to recalculate the composition of the carbonate on a carbon dioxide-free basis. This is done using the appropriate relative atomic masses (atomic weights), as outlined in the example below, which illustrates how to plot the composition of dolomite in the system SiO_2–CaO–MgO.

Dolomite has the general formula $CaMg(CO_3)_2$, which can be recast as $CaO \cdot MgO \cdot 2CO_2$. Its relative molecular mass (molecular weight) is 184 g, obtained by summing the relative atomic masses (which have been rounded to the nearest whole number) as follows:

	Relative atomic mass	Number of atoms	Product
Ca	40	1	40
Mg	24	1	24
C	12	2	24
O	16	6	96
Sum		10	184

The proportions of CaO and MgO within dolomite are given by the ratios of the relative molecular masses of CaO (56) and MgO (40) to 184, and are 30.4% and 21.7%, respectively, totalling 52% and leaving 48% CO_2, which is lost to the atmosphere on firing. Thus the relative proportions of CaO and MgO are 30.4/52 = 58% and 21.7/52 = 42%, respectively, and so mark the position of dolomite when plotted in Figure 5.10.

5.4 TIME–TEMPERATURE–TRANSFORMATION (TTT) DIAGRAMS

Phase diagrams, such as those introduced above, represent the mineral (and melt) phase relationships for systems which are at chemical equilibrium. In many industrial processes, chemical equilibrium may be energetically an expensive luxury. It may require long periods of heating to produce an equilibrium assemblage, when a suitable product might result from a shorter period of heating to produce a mineral assemblage which has the desired properties, but which need not have reached equilibrium. To understand the way in which reactions approach equilibrium the effect of time has to be considered, and a convenient but rather labour intensive approach has been used by Dunham and Scott and their coworkers (e.g. Dunham, 1992; Kimyongur and Scott, 1986). TTT diagrams are compiled empirically by experimentally firing the raw material of interest at a number of different temperatures for a number of different times, quenching and identifying the mineral products. In this way the minimum amount of time or the minimum temperature required to produce a given mineral product can be determined, with potential savings in the amount of energy required to fire the material. This approach works best for pure minerals, such as carbonates or simple clays, where compositional varia-

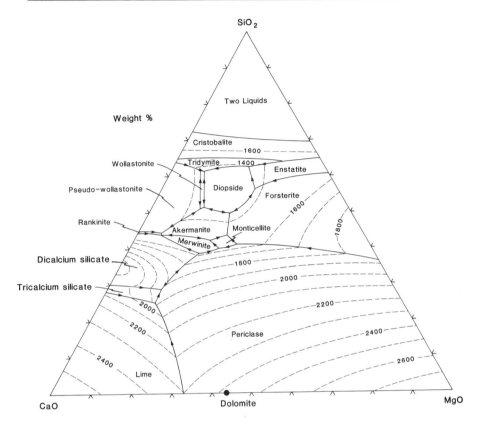

Figure 5.10 Phase diagram for the three-component (ternary) system SiO_2–MgO–CaO, at atmospheric pressure, showing the liquidus fields for minerals (and artificial compounds) containing Mg, Ca, Si and O and the composition of dolomite. This diagram is plotted with compositions expressed as weight %; temperatures in °C.

tion can be neglected and where the stoichiometry of the phase changes is straightforward. An example of a TTT diagram, for magnesite, is given in Figure 5.11a. This figure shows the matrix of times and temperatures that were used, together with the the identified mineral products. The firing of magnesite follows a simple decarbonation or calcining reaction:

$$MgCO_3 = MgO + CO_2 \qquad (5.6)$$
$$\text{magnesite} \quad \text{periclase} \quad \text{carbon dioxide}$$

Figure 5.11b illustrates the potential value of TTT diagrams, as it shows variation in a measured property, the surface area of the calcined product, as a function of temperature and heating time. The contoured values show that a maximum surface area is produced at low temperatures within the

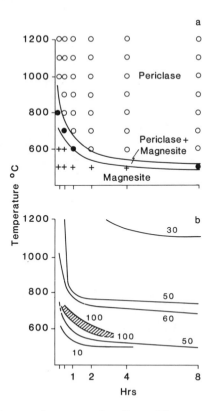

Figure 5.11 TTT diagram for magnesite (from Kimyongur and Scott, 1986), showing (a) the mineral phases present as a function of heating time and temperature and (b) the specific surface areas of the heated products, contoured in m^2/g.

periclase stability field, and for low heating times. Thus the periclase produced by calcining is initially very fine grained, with a high surface area, and coarsens with reducing surface area as heating is prolonged or as the temperature is raised. The reactivity of the calcined periclase is proportional to its surface area, and so it is at its most reactive when produced under the optimum conditions defined by the shaded time and temperature field in Figure 5.11b. In this case, overheating might do more harm than good, as well as wasting energy. Similar diagrams can be produced for other minerals or mineral mixtures.

5.5 NOTATION AND CONVENTIONS

In many ceramic applications, particularly cement manufacture, it is usual to abbreviate chemical formulae to a shorthand notation, which facilitates

NOTATION AND CONVENTIONS

writing chemical formulae (especially those which are complex or for compounds for which there is no mineral name). The abbreviations that are commonly used are as follows:

- CaO = C
- SiO_2 = S
- Al_2O_3 = A
- Fe_2O_3 = F
- MgO = M
- Na_2O = N
- K_2O = K
- H_2O = H
- CO_2 = $\bar{C}$
- SO_3 = $\bar{S}$

You will also have noticed that in some of the formulae used in the figures certain elements are grouped in parentheses, separated by commas as in the following example for lime: (Ca,Mg)O. This tells the reader that Ca and Mg are able to substitute for one another by solid solution, and that their proportions are not necessarily fixed. The compositional limits to solid solutions, in this case determination of the amount of Mg that can substitute for Ca in lime, are usually determined by analysis of natural or synthetic crystalline phases. It should also be noted that most phase diagrams are based on weight proportions rather than atomic proportions, allowing direct use of analytical data. However, if uncertain or if a diagram has been inadequately labelled, a user should always check by plotting a mineral phase of known composition.

6 Raw materials for the glass industry

Approximately three millions of tonnes of glass are produced each year in the United Kingdom, 21 million tonnes in the European Union altogether, of which approximately 66% is made as containers and 22% produced as flat or window glass. The USA produces approximately 15 million tonnes of glass annually and Japan 13 million tonnes. The world leader in flat glass production is the British company Pilkingtons of St Helens, which was responsible for invention of the float glass process described below. In addition to glass appropriate for the mass markets of windows and containers, there are several specialist glasses designed for specific purposes. Perhaps the most commonly encountered of these are glasses which are resistant to thermal stress, and which are found in the domestic kitchen in cooking utensils and hobs. Glass manufacture is presently a sophisticated business, with many new opportunities for a material with a long history which is widely taken for granted.

In terms of quantity, most of the glass which is produced today is designed for use as containers (bottles and jars etc.) or as windows. This type of glass is known as a soda–lime–silica glass, reflecting its simple ingredients, obtained from the raw materials shown in Table 6.1.

The composition of the glass lacks carbon dioxide, which is lost as the carbonate raw materials decompose. Soda–lime–silica glasses have been

Table 6.1 Raw materials used in lime–soda glass

Raw materials	approximate proportion (wt %)	provides	approximate proportion in glass (wt %)
soda ash (Na_2CO_3)	25	soda (Na_2O)	18
limestone ($CaCO_3$)	10	lime (CaO)	7
silica sand (SiO_2)	65	silica (SiO_2)	75

Table 6.2 Examples of glass compositions (wt %; from Austin, 1984)

Component	1	2	3	4	5	6	7	8	9
SiO_2	67.8	69.4	72.0	72.4	81	72.4	67.2	96.3	55
B_2O_3	–	–	–	–	12.5	–	–	2.9	10
Al_2O_3	4.4	3.5	2.1	0.8	2.0	1.0	–	0.4	14
Fe_2O_3	–	1.1	–	0.4	–	0.1	–	–	–
As_2O_3	–	–	–	–	–	–	0.5	–	–
CaO	4.0	7.2	10.2	5.3	–	8.1	0.9	–	13
MgO	2.3	–	–	3.7	–	0.2	–	–	5
Na_2O	13.7	17.3	14.9	17.4	4.5	18.1	9.5	–	0.5
K_2O	2.3	–	–	–	–	–	7.1	–	–
PbO	–	–	–	–	–	–	14.8	–	–
BaO	–	–	–	–	–	0.2	–	–	–
Li_2O	–	–	–	–	1.0	–	–	–	–
SO_3	1.0	–	0.8	–	–	–	–	–	–

1, Egyptian glass (1500 BC); 2, Roman glass (around 290 AD); 3, soda–lime–silica container glass; 4, electric lamp bulb glass; 5, Pyrex laboratory glassware; 6, soda–lime–silica glass tableware; 7, lead crystal tableware; 8, Vycor high temperature scientific glassware; 9, 'E' glass for fibres

produced since Roman times, with only minor subsequent changes in their formulation (Table 6.2). Other glasses, designed to meet particular technical specifications, may be based on compositions which include boric oxide (to produce heat-resistant glasses such as 'Pyrex' and 'Vycor') and lead oxide (for lead crystal tableware). Alumina is a widespread component of glasses in addition to soda ash and silica, and helps improve resistance to weathering. Magnesium can be substituted for a proportion of the calcium content by the use of dolomite instead of limestone, and potassium can be substituted for some of the sodium with the use of feldspar, aplite or nepheline syenite. Opaque glasses are produced by the introduction of fluorides. Non-silicate glasses are becoming increasingly important for special optical purposes, for example in the use of glasses prepared from CaF_2, AlF_3 and P_2O_5 for infrared optics or the use of fluoride glasses for optical fibres.

Great care is taken to consider the minor components of a glass, as small traces of impurities may have a major positive or negative effect on the quality of the finished product. For example, the presence of traces of iron may give a pale green colour (often visible when examining a pane of glass end on), and this can be tolerated in some applications (such as container glass). Other minor components might have beneficial effects on the qualities of the glass produced. For example, addition of lithium reduces the temperature required to melt the glass, and so yields savings in energy costs (against which has to be offset an increased cost of the raw material). Other components are added to influence the refractive index or other physical properties of the glass. Coloured glasses are produced by addi-

Table 6.3 Composition of glass raw materials (wt %; from Lefond, 1983)

Component	White sand	Yellow sand	Lime stone	Dolomite	Feldspar	Aplite	Nepheline syenite	Blast furnace slag
SiO_2	99.7	99.1	0.7	0.5	68.0	60.5	59.5	38.0
Al_2O_3	0.08	0.25	0.1	0.3	19.4	23.0	23.9	7.4
Fe_2O_3	0.025	0.15	0.075	0.075	0.075	0.41	0.085	0.30
Cr_2O_3	0.0002	0.002	0.001	0.0005	0.0003	0.0003	0.0003	0.0030
TiO_2	0.02	0.05	–	–	0.01	0.20	–	0.40
CaO	0.01	0.01	54.7	31.0	1.4	5.6	0.2	38.8
MgO	0.01	0.01	0.60	21.0	0.02	0.05	–	12.5
Na_2O	0.004	0.015	–	–	6.9	6.1	10.5	0.4
K_2O	0.005	0.050	–	–	4.0	2.9	5.0	0.4

tions of small amounts of the colouring agents, such as Fe (green), Ni (brown) and Co (blue). In detail, the colour of the glass depends on the bulk composition and how the trace element is accommodated into the melt structure: for example, NiO dissolved in a soda–lead glass gives a brown colour, but an orange colour in a potassic glass. Components such as Cr give glasses which vary in colour according to the valency of the chromium. Colour can also be introduced by the incorporation of solid inclusions, either introduced with the raw materials or by the growth of crystals within the glass.

As tiny amounts of impurities can greatly affect the quality of the finished glass, one of the first requirements for the raw materials is that their purity should be known, so that the final mixes of raw materials can compensate for natural variation in the quality of each. Examples of raw materials, together with typical compositions, are given in Table 6.3. Note that the amounts of chromium and iron in particular are tightly specified.

6.1 GLASS MANUFACTURE

In general terms, soda–lime–silica glass manufacture involves melting the required raw material mix at 1600°C, which yields a very fluid melt, from which gases can escape (especially carbon dioxide produced by the decomposition of carbonate raw materials). The glass is then worked to produce the articles required at about 1000°C, followed by annealing at 500–600°C. The float glass process, used to produce flat panes of glass suitable for windows, illustrates this well (Figure 6.1). Before the invention by Pilkingtons of the float glass process, window glass was produced either by smoothing and polishing mechanically flattened or spun pieces of glass in a very labour-intensive and hazardous process, or by drawing a ribbon of glass vertically from the melting chamber (Austin, 1984). In the float glass

GLASS MANUFACTURE

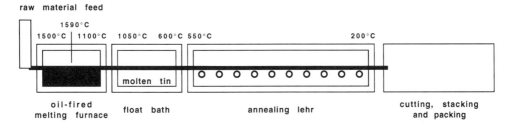

Figure 6.1 Diagram of the float glass process, showing the way a continuous ribbon of glass is drawn from the melting furnace, through the float bath (which gives the perfect surface to the sheet) and then is annealed and allowed to cool before preparation for sale.

process, the flat sheet of glass is allowed to form by floating on a bed of molten tin, under a reducing atmosphere. Thus the surfaces of the glass sheet are formed against a liquid and a gas and are perfectly smooth, requiring no subsequent polishing. The thickness of the glass sheet is controlled by varying the rate at which the molten glass is withdrawn from the melting vessel and, after annealing, sheets of the required area are cut automatically from the moving ribbon of glass. Containers and other shaped products are made by blowing or pressing into moulds at the working stage. Glass fibres are spun from a rotating perforated disc, and both glass and optical fibres are drawn from fine nozzles.

6.1.1 The chemistry of glass manufacture

A glass is little more than a rapidly quenched liquid, which behaves as a solid but retains the molecular structure of the liquid. The term 'glass' can be applied to many different materials, but in common usage it refers to quenched silicate liquids. The production of commercial glasses is therefore dictated by the application of phase diagrams such as those introduced in Chapter 5, which allow the melting behaviour of particular compositions to be predicted and the optimum conditions for glass manufacture to be identified. For common window or container glass, the appropriate phase diagram is that for the system SiO_2–CaO–Na_2O (Figure 6.2).

In detail, liquidus phase relationships within the three-component system SiO_2–CaO–Na_2O go well beyond those relevant for glass manufacture. Consequently, Figure 6.2 focuses on the silica-rich corner of the triangular diagram, as this includes most glass compositions. In this region, the silica mineral on the liquidus is cristobalite, tridymite or quartz (depending on temperature), with very steep temperature gradients particularly towards more sodic compositions. Towards the lime apex, a field of two liquids is drawn; in this field, liquid compositions separate out into two contrasting liquids, one silica-rich and one lime-rich. These two liquids

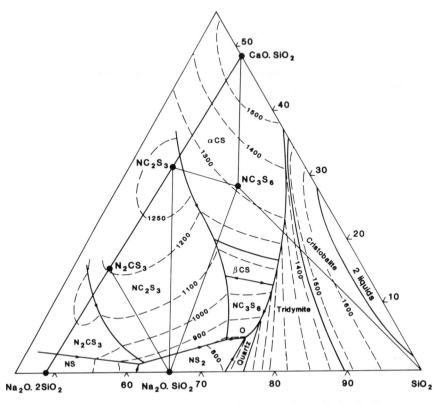

Figure 6.2 Phase relationships for part of the system SiO_2–CaO–Na_2O at atmospheric pressure (weight %). Point O is the ternary eutectic, at 725°C, with the composition 5.2% CaO, 21.3% Na_2O and 73.5% SiO_2. Compositions are expressed as weight %. The system includes the following crystalline phases:

Name	Formula	Abbreviated formula
Cristobalite	SiO_2	S
Tridymite	SiO_2	S
Quartz	SiO_2	S
Pseudowollastonite	$CaSiO_3$	CS
Sodium silicate	Na_2SiO_3	NS
Sodium disilicate	$Na_2Si_2O_5$	NS_2
Sodium calcium silicate	$Na_4CaSi_3O_9$	N_2CS_3
Sodium calcium silicate	$Na_2Ca_2Si_3O_9$	NC_2S_3
Sodium calcium silicate	$Na_2Ca_3Si_6O_{16}$	NC_3S_6

are immiscible in the same way that oil and water are immiscible, and like a good mayonnaise they are opaque to light and can be quenched to produce an opaque white solid. The other liquidus fields show shallower temperature gradients. On the boundaries between them arrows are marked to show the 'downhill direction'. These all converge on a single point, where the temperature at which liquid can exist is lowest, which is a ternary eutectic. The ternary eutectic composition can be read from the compositional axes and corresponds to 5% CaO, 21% Na_2O and 74% SiO_2. The minimum temperature can be read from the contours (or in most published diagrams is given more precisely in accompanying tables), and is 725°C.

In order to decide on the optimum blend of ingredients required to make a soda–lime–silica glass, the ternary liquidus diagram can be used to indicate the temperature required to initiate melting. The ternary eutectic composition is therefore the one which appears to be ideal for glass manufacture, as it will begin to melt at the lowest temperature, saving energy and manufacturing costs. Compositions close to this have been used since Roman times, but recent practice involves compositions which are slightly richer in silica. Melting is carried out at 1600°C to give enough superheat to ensure that all of the solid grains within the raw materials dissolve within the liquid, and to ensure that the viscosity of the liquid is sufficiently low that gases can escape. Compositions which are more silica-rich have a rapidly rising liquidus temperature, and may not completely melt, leaving a glass which contains crystals of a silica mineral or bubbles and appears frosted. It is therefore important to use this and similar diagrams not only to design batch mixes but also to diagnose problems which arise when glasses are not correctly made.

The sources of soda and lime are respectively sodium carbonate (soda ash) and limestone (dolomite is used if magnesium is needed). These materials decompose on heating with the loss of carbon dioxide. This is desirable in glass manufacture, where one thinks in terms of simple oxide components of the glass, but it results in the batch having to degas as it is heated. The compositions and temperatures used have to be designed to provide an appropriate combination of melt properties, particularly viscosity, to allow the escape of carbon dioxide and any other gases that may be present, such as water. Thus, in the formulation of batches consisting primarily of silica sand, limestone and soda ash, proportions must be corrected to take into account the loss of carbon dioxide so that they correspond to the compositions required for the finished glass. In order to carry out this correction, relative atomic masses (atomic weights) are used to determine the proportions of CaO within $CaCO_3$ and Na_2O in Na_2CO_3:

- Relative atomic masses: $Ca = 12; O = 16; Na = 23; Ca = 40$
- Relative molecular masses: $CaO = 56; Na_2O = 62; CO_2 = 44$
  $CaCO_3 = 100; Na_2CO_3 = 106$

Therefore 100 tonnes of limestone ($CaCO_3$) yields 56 tonnes of CaO and 44 tonnes of CO_2, and 100 tonnes of soda ash (Na_2CO_3) yields $100 \times 62/106 = 58$ tonnes of soda and 42 tonnes of CO_2. The carbon dioxide evolved in this way escapes to the atmosphere, making a not insignificant contribution to anthropogenic emissions.

6.2 GEOLOGY OF GLASS RAW MATERIALS

The raw oxides Na_2O and CaO that are components of the phase diagram used to discuss glass manufacture (Figure 6.2) do not occur naturally as stable mineable commodities. Instead, the corresponding carbonates are used, taking advantage of the fact that carbon dioxide will be lost during the melting process, and silica sand is used as the source of SiO_2. Having discussed the origin of sodium carbonate in Chapter 4, we will now examine the geological factors which control the occurrence and quality of silica sand and limestone suitable for glass manufacture. As shown in Tables 6.2 and 6.3, glass compositions are variable and can be achieved by blending raw materials in a number of possible combinations. Impurities in one raw material can be tolerated, but there may need to be compensation by reduction in the amount of another raw material. An example of this can be provided by the use of albite ($NaAlSi_3O_8$) as a source of Al to increase the resistance of flat glass to weathering; the addition of albite involves addition of Na and Si as well as Al, and so in compensation the amounts of soda ash or silica sand needed to achieve a particular bulk composition are reduced.

6.2.1 Silica sand

Sand is of course one of the commonest geological materials, as unconsolidated or consolidated sediment. Silica sand suitable for glass manufacture is however relatively rare, because of the need for a high degree of chemical purity. The 'holy grail' for silica sand producers is a sand that can achieve a silica content of 99.99% SiO_2 after beneficiation.

The essential requirements for silica sand for glass manufacture are that it must be even grain size – more than 90% of grains must lie in the range 125–500µm, and its chemical composition must meet the requirements shown in Table 6.4.

Table 6.4 Required chemical composition of silica sand for glass manufacture

Glass	Minimum SiO_2	Maximum Fe_2O_3	Maximum Cr_2O_3
Opthalmic glass	99.7	0.013	0.00015
Tableware, crystal and borosilicate glass	99.6	0.010	0.0002
Colourless containers	98.8	0.030	0.0005
Coloured containers	97.0	0.25	–
Clear flat glass	99.0	0.10	0.0001

The discolouring impurities iron and chromium occur within the non-quartz mineral fraction of the sands. Iron can occur as haematite, giving the sand a red colour, or as oxy-hydroxides (giving a yellow or brown colour) as well as in silicate minerals. Chromium occurs as the heavy mineral chromite ($FeCr_2O_4$), which is stable during weathering and erosion and is ultimately derived from basic or ultrabasic igneous rocks. Chromite is also stable during glass manufacture, and so rather than resulting in a discoloured glass, it persists as solid inclusions within the finished product, which can cause it to be brittle. This is especially important for float glass manufacture, where persistence of chromite grains can render useless substantial lengths of glass strip. Because of the difficulties involved in the chemical determination of minor amounts of Cr it may be appropriate simply to count the number of grains of chromite detected optically within a sample of known weight in order to classify a sand as suitable for float glass.

Alumina is a natural impurity in glass sands, arising from the presence of feldspars, mica or clay minerals, and varies from 0.4% to 1.2% Al_2O_3. High values in this compositional range are preferred because they help to reduce melting temperatures (yet another component is added) and involve no negative effect on glass colour or other physical properties. The occurrence of aluminium as an impurity may also be beneficial by reducing the need to add aluminosilicates (feldspar, aplite or nepheline syenite) for the manufacture of certain glasses.

The physical specifications relate to the need to produce glass by melting a reactive mixture of solids. An even size of the quartz grains facilitates melting – put simply, unevenly large grains might not completely melt in the time available. Fine grains melt most rapidly because of their relatively large surface area giving a high reactivity, and because of an increased number of grain contacts. In addition, the removal of material with a grain size below that specified, by washing the sand, removes clay minerals which introduce chemical impurities. Of major importance is consistency of supply over a period of several years, as the raw materials enter an industrial process which is not necessarily adaptable to changing conditions.

Bearing in mind the constraints imposed by these technical requirements, the production of a glass sand depends very much on the availability of pure quartz sands. Sands of sufficient purity to supply the most demanding customers are rare. In Britain, only the Cretaceous Loch Aline sands are suitable for high grade tableware and laboratory ware, achieving 99.8% SiO_2 (Table 6.5). The value of these sands justifies underground mining of a bed 10 m thick which contains a pure white, poorly cemented, unit 3–7.5m thick. Quaternary sands, especially those of Cheshire and Lancashire, are an important source for lower grades of sand, supplying the float and container glass industries. Elsewhere in Europe, the most

Table 6.5 Typical compositions of glass sands and a foundry sand (wt %; Highley 1977)

Component	1	2	3	4	5	6
SiO_2	99.80	99.70	97.5	96.8	95.1	97.0
Al_2O_3	0.08	0.12	1.3	1.5	2.25	1.63
Fe_2O_3	0.011	0.026	0.105	0.125	0.35	0.15
Cr_2O_3	0.0003	0.0004	0.001	–	–	–
TiO_2	0.01	0.02	–	–	0.11	0.06
CaO	0.01	0.01	0.1	–	0.18	0.11
MgO	0.01	0.01	–	–	0.13	–
Na_2O	0.01	0.01	0.1	0.2	0.28	0.14
K_2O	0.005	0.01	0.6	0.8	1.1	0.74

1, Loch Aline Grade B (colourless tableware and laboratory glass); 2, Loch Aline Grade C (colourless tableware and laboratory glass); 3, Pleistocene white sand, Cheshire (float glass); 4, Recent wind blown sand, Lancashire (float glass); 5, Recent wind blown sand, Lancashire (coloured glass; as dug); 6, Pleistocene brown sand, Cheshire (foundry sand).

famous glass sands are those of Fontainebleau, south of Paris, which are remarkably pure silica sands; Belgium is also a major producer. In North America, the Ordovician St Peter Sandstone (producing the 'Ottowa' sand) and the Devonian Oriskany Sandstone are worked at locations where a diagenetic cement is lacking or easily removed by washing. Sands which naturally reach the specifications required tend to be very rare, and even high quality sands may require beneficiation by washing. If produced from sandstones these must lack a cement, unless easily removed by washing, to avoid any need to grind the rock. When seen in the field, glass sands are striking in their brilliant whiteness. Sands which fail to meet the requirements of the glass industry may be sold for other applications, especially as foundry sands used for preparing moulds for metal casting.

(a) The Cheshire silica sands

In Cheshire, silica sands are worked from deposits of the Pleistocene Chelford Formation (Figure 6.3). They rest directly on Triassic rocks or on glacial till (boulder clay), and are overlain by either an impure sand (the Gawsworth Formation) or till. The sands are believed to be derived from very pure sandstones within the Namurian (Millstone Grit) sequence which outcrops to the north, which have also been quarried for other purposes. The Chelford Formation consists of a pure white glass sand (Table 6.5) which lies beneath a less pure brown sand suitable for use in foundry work – it has a higher iron content, and contains smectite clay impurities which act as a binder when making moulds for iron casting. The Chelford glass sand requires processing as follows:

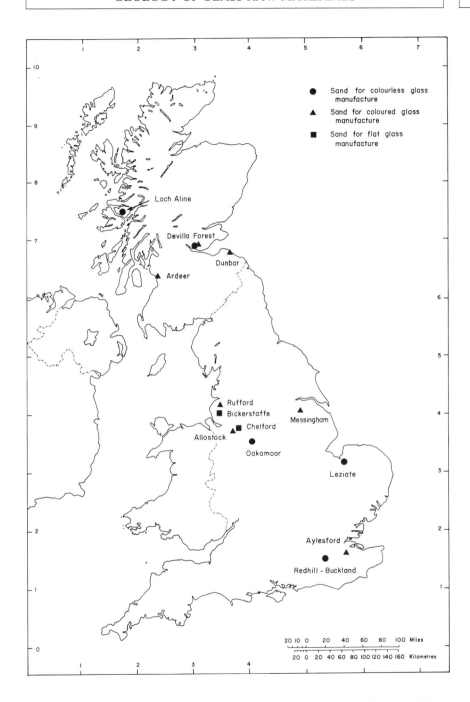

Figure 6.3 Sources in the United Kingdom of glass sands in 1976 (from Highley, 1977).

- screening
- washing and sizing, using hydrocyclones, attrition scrubbing etc.
- flotation and/or chemical/acid leaching to remove Fe minerals and stained quartz
- spiralling and/or tabling to remove heavy minerals
- hydrosizing
- dewatering

The deposits are worked by selective mining in open pits, controlled by an extensive quality evaluation exercise which is carried out prior to mining. The compositions of the sands are determined by X-ray fluorescence analysis of samples taken from boreholes ahead of the working face, up to one year in advance of mining. Both foundry sand and glass sand can be produced from a given quarry. Glass sands are sold wet whereas foundry sands are dried before sale, leading to a differential in price (1993 values):

- glass sand – £10–15 per tonne
- foundry sand – £12–17 per tonne
- compared with building sand – £5 per tonne washed and graded, £2–3 per tonne as dug.

6.2.2 Limestone

Limestone is required twice in glass manufacture – once to produce sodium carbonate (see Chapter 4), and secondly as an ingredient in the batch to be melted. As an ingredient in batches to be melted to produce glass, limestone purity is critical. In particular, Fe contents have to be very low, and the amount of MgO, as in dolomite, has to be known. In some glasses MgO is added using pure dolomite, but the amounts have to be controlled. Like CaO, MgO causes immiscibility in glass melts; the miscibility gap in the system SiO_2–MgO is wider than that in the system SiO_2–CaO. Pure limestone suitable for glass manufacture is relatively rare within the UK – it is worked from the Carboniferous of Derbyshire (particularly the Bee Low Limestone at Buxton), underground near Wirksworth, at Dove Holes and elsewhere. Most dolomite in Britain is produced from the Permian Magnesian Limestones, which outcrop in a strip extending from Nottingham through Yorkshire until it meets the coast of County Durham near Sunderland, and again one of the main requirements here is for the iron content to be very low.

Limestone is such a widespread rock that it would be foolish to attempt to discuss here all aspects of limestone geology. Instead, emphasis will be placed on those factors which distinguish deposits of limestone which are suitable for glass manufacture from those which are not. It is important that in any investigation of a limestone, consideration is given to achieving the best price for a quarry product, and so the possibility of producing limestone for the chemicals or glass industries should be ruled out before settling for production of a lower specification (and hence lower price) material such as aggregate. The proportion of limestone that can be sold as the most highly specified product should be maximized.

Limestone deposits occur predominantly as bedded units with considerable lateral extent, and also as reefs representing local accumulations which originated as bioherms (such as coral or algal reefs) or current-sorted clastic material. As in any sedimentary environment, accumulation depends on the nature of the material supplied to the depositional environment. Clastic or volcanic material may contaminate the limestone, either as a dispersed phase or as interbedded units. Limestones can be thickly bedded, with individual bedded units several tens of metres thick, or thinly interbedded with clays or other clastic rock on a scale of tens of centimetres or less. Reefs can represent sources of high quality limestone irregularly surrounded by poorer material. After deposition, diagenetic processes may redistribute some constituents, and alteration associated with mineralization may introduce new contaminants. Geologically-produced variation in chemical purity arises largely from the dilution of calcium carbonate (the essential component of pure limestone) by other material:

- primary depositional impurities include grains of clay or other clastic minerals incorporated into the limestone; these introduce silica, alumina, iron and other chemical impurities.
- volcanic activity may result in the introduction of ash as a dispersed component of the limestone, or in the formation of beds of volcanic ash composed of clay minerals and residual silicates (these are often bentonites, but too thin to be economically interesting; Chapter 3).
- diagenetic processes affecting silica can result in the formation of chert or flint as nodules which may connect to form irregular layers parallel to bedding.
- diagenetic dolomitization can result in the massive transformation of limestones to form dolomite, with the formation of the mineral dolomite $CaMg(CO_3)_2$ in place of calcite and destruction of primary depositional textures. Taken to extremes, dolomitization can result in the complete transformation of limestone to produce dolomite rock, which may be valuable as a source of dolomite.
- hydrothermal alteration associated with mineralization (limestones are characteristically host to Mississippi Valley Type mineral deposits) can produce irregular dispersion of ore-forming elements, including fluorine, adjacent to mineral veins and faults. The presence of fluorine is particularly undesirable for limestones which are to be fired, as it escapes with the flue gases to form hydrofluoric acid during rainfall.

- contact metamorphism, especially of impure limestones, adjacent to igneous intrusions can result in the formation of calcium silicate minerals dispersed throughout the rock.

As an example of the assessment of a limestone province carried out in part to identify occurrences of high quality deposits, some of the results of a strategic survey of limestones within the Peak District of England (Harrison and Adlam, 1985) are summarized in Figures 6.4–6.7. This area is the largest outcrop of Carboniferous limestones in England, and is heavily quarried (Figure 6.4), despite lying largely within a national park. Detailed chemical analysis has enabled the area to be mapped according to limestone quality, where the purity of the limestone has been estimated on the basis of its calcium carbonate content (Figure 6.5). This assessment is reinforced by the mapped distribution of impurities such as silica (Figure 6.6), which shows highest values within impure limestones, characterized by thinner bedding with clay-rich or cherty facies. Figure 6.7 shows the widespread suitability of the limestones of the province for aggregates, which are less demanding in terms of their chemical composition.

From the point of view of production of limestones for glass manufacture it is essential that the consequences of the impurities listed above are minimized. Forward drilling ahead of working faces is essential to determine the frequency and nature of variation in the quality of the limestone, combining logging of geological features such as mineral veins and bedding with chemical analysis of samples taken from boreholes. As with sands, consistency in supply over long periods is required to satisfy the industrial customers for a quarry's products.

6.3 MINOR CONSTITUENTS: THE USE OF LITHIUM IN GLASS MANUFACTURE

In terms of the major constituents of common glass, most industrial countries are self sufficient. Minor constituents can however be extremely rare. As an example, we will consider lithium, which is a valuable additive to glass compositions (Ansems, 1990; Kingsnorth, 1988). Additionally, lithium occurs in a number of mineral forms, and the use of lithium in glass manufacture has created new markets for some lithium minerals.

Lithium is added to glasses for several reasons, because it reduces liquidus temperatures; it improves moulding properties (reduces viscosity); it improves thermal properties ('Pyrex', ceramic hobs) and it improves strength. The amounts required are very small, frequently less than 1%, and not more than 4%.

These benefits have usually been developed using lithium carbonate as a source of lithium, with the advantage that only Li_2O is added to the glass

MINOR CONSTITUENTS: LITHIUM IN GLASS MANUFACTURE

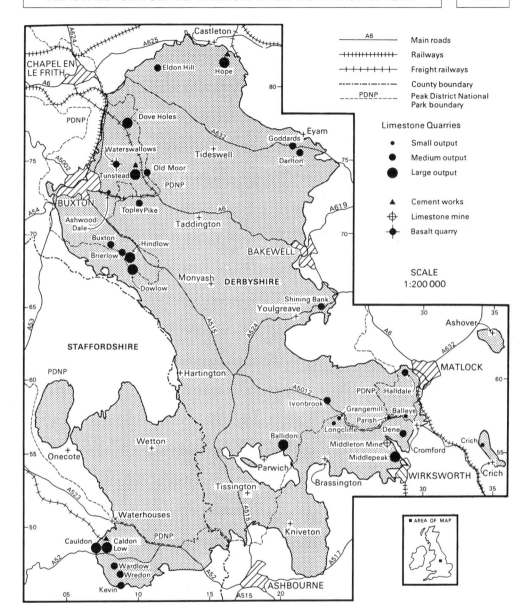

Figure 6.4 Location of working limestone quarries in the Peak District of Derbyshire (early 1980s; from Harrison and Adlam, 1985).

composition as CO_2 is lost to the atmosphere. However, lithium carbonate is not a cheap chemical – it is produced from brines, or manufactured from the lithium minerals amblygonite ($LiAlPO_4(F,OH)$), petalite ($LiAlSi_4O_{10}$) and spodumene ($LiAlSi_2O_6$). These three minerals are rare, but occur in

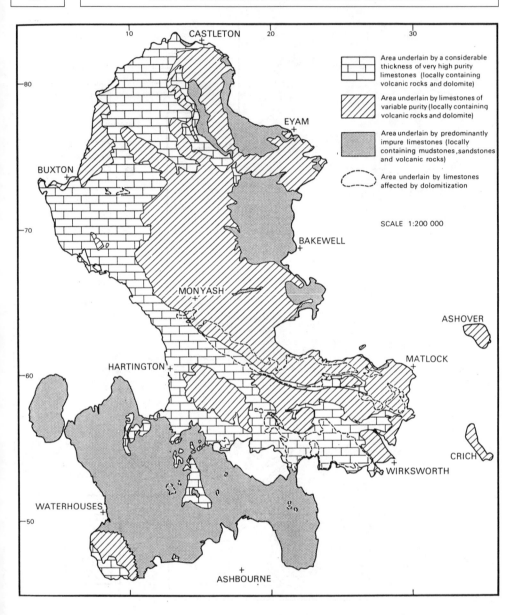

Figure 6.5 Summary of limestone quality for the Peak District (from Harrison and Adlam, 1985).

concentration in a number of places as constituents of pegmatites – the rare metal pegmatites. Rare metal pegmatites are mineralogically complex, containing a diverse variety of mineral species, including many minerals which are host to rare metals such as Li, Ta, Nb and Cs. In addition to the

MINOR CONSTITUENTS: LITHIUM IN GLASS MANUFACTURE

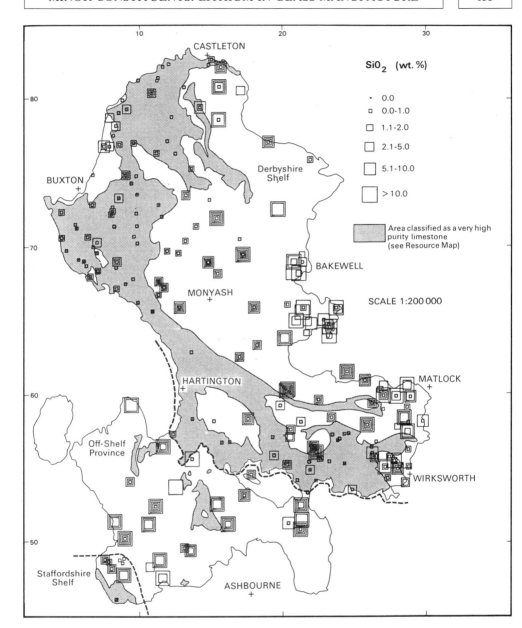

Figure 6.6 Silica contents of Peak District limestones (from Harrison and Adlam, 1985).

Li minerals, feldspars and quartz occur, together with lithium micas (not currently of commercial interest), and the Cs–Nb–Ta minerals may represent a coproduct or by-product. The mineralogy of the rare metal

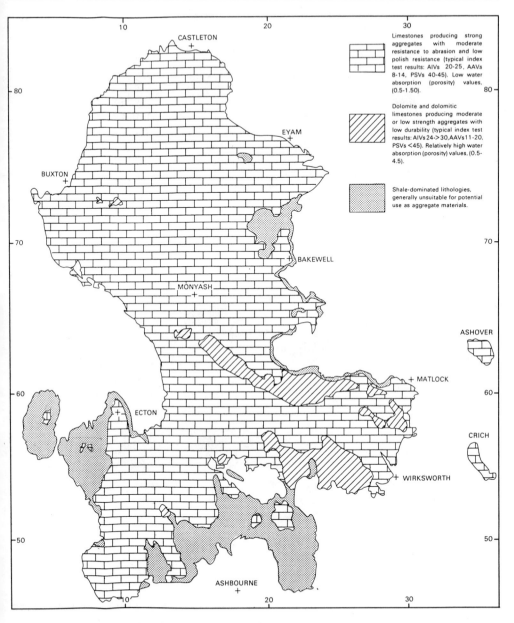

Figure 6.7 Suitability of Peak District limestone for aggregate use (from Harrison and Adlam, 1985).

pegmatites is exotic, to say the least, and typically shows zoning (Figure 6.8). Fortunately, as pegmatites they are coarse grained, sometimes spectacularly so, and large individual mineral grains are often pure and

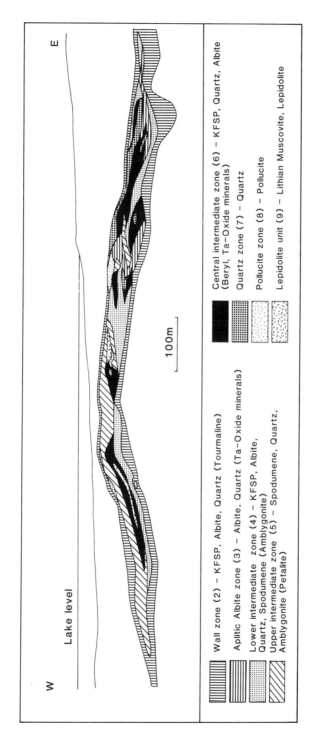

Figure 6.8 Mineralogical zoning in the Tanco pegmatite (from Crouse et al., 1979).

easily separated during selective mining. Iron-bearing minerals are sparse – the rare metal pegmatites are dominated by alkali alumino-silicate phases.

Localities where rare metal pegmatites are worked are limited to the Tanco pegmatite (Manitoba, Canada; Burt, Fleming, Simard *et al.*, 1988), pegmatites at Greenbushes (Queensland, Australia) and those of North Carolina (USA), with dormant or less commercially significant deposits in Zimbabwe (Bikita) and south-west Africa (Namibia). The first three supply the bulk of the industrial world's requirements. In addition, lithium pegmatites are also known to occur in south-east Ireland (adjacent to the Leinster granite), Devon (the Meldon aplite is mineralogically analogous), Spain and France, and have been recently reported in Austria. Supply theoretically exceeds demand, because very small amounts are required for glass manufacture, and these ores cannot yet be used to economically produce Li for metallurgical purposes. Initially, many pegmatites of this type were investigated as sources for the rare metals, tantalum, niobium and rare earths, regarding the predominant lithium aluminium silicates as gangue or waste. However, the market for rare metals has proved too weak to justify their sole production from many of deposits, and the lithium aluminosilicates may now be the major justification for mining, with the rare metals as by-products.

The origin of the rare metal pegmatites is not well understood. They may be the products of extreme fractionation of granitic magmas, but often are not temporally associated with granitic rocks. Those which occur within high grade metamorphic terrains may well represent anatectic melts of peculiar composition, perhaps reflecting very small amounts of partial melting. They are believed to have crystallized at low temperatures, around 400–500°C at 3 Kb to 600°C for volcanic equivalents, which have been recognized in Peru. Fluid inclusion evidence shows that their crystallization was in the presence of a highly saline lithium- and boron-rich fluid: although boron minerals are rare in Li pegmatites, they commonly occur within associated metasomatic aureoles.

The stability of the lithium aluminium silicate minerals has now been determined, and is summarized in the phase diagram shown in Figure 6.9.

One of their characteristics is that petalite is often the primary mineral to form, and as it cools it reacts to form a spodumene-quartz intergrowth, known as 'squi', according to the reaction:

$$LiAlSi_4O_{10} = LiAlSi_2O_6 + 2SiO_2 \tag{6.1}$$

petalite = spodumene + quartz

Cooling pathways for three examples of lithium pegmatites are shown in Figure 6.9, explaining the occurrence of this phenomenon in the Tanco and

MINOR CONSTITUENTS: LITHIUM IN GLASS MANUFACTURE

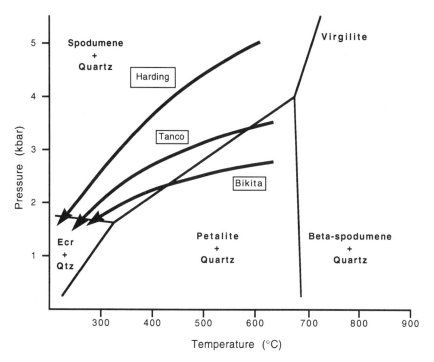

Figure 6.9 Stabilities of Li-Al-silicate minerals as a function of temperature and pressure (adapted from London, 1984). Ecr = eucryptite and Qtz = quartz. The arrows indicate cooling pathways deduced from the mineral assemblages preserved at the Bikita, Tanco and Harding pegmatites.

Bikita pegmatites, but not the Harding pegmatite which crystallized out at higher pressures than the other two (London, 1984). This reaction has consequences for mineral processing and beneficiation, because the liberation of spodumene from squi is a far from trivial problem. However, as far as glass manufacture is concerned this problem can be neglected. Squi can be used without further treatment for the purposes of glass manufacture as it is derived from petalite, which is chemically identical to a spodumene–quartz mixture, and is known as 'glass grade spodumene'. There are no more detrimental impurities than there would be in pure petalite and spodumene, although some correction might be required for the increased proportion of silica if squi is used instead of spodumene. In addition to adding Li to the glass, the use of lithium aluminosilicates involves the addition of aluminium. However, the amounts involved are small: 1% Li_2O (which is a lot) represents 12% spodumene in the batch; this adds about 2% Al_2O_3, which for many glasses would be added from other sources in any case.

6.4 SOME ENVIRONMENTAL ASPECTS OF GLASS MANUFACTURE

Glass is a material essential to everyday life, and it is usually used with little thought to the environmental implications of its production. In recent years, however, it has become normal to recycle domestic glass waste, and the general public clearly identifies the recycling of glass as an environmentally sound practice. It is usually assumed that the major saving introduced in this way lies in the reduced consumption of the sand, soda ash and limestone raw materials, but there are other considerable benefits. First, the use of scrap glass is long established in the industry as a means of reducing energy costs – if glass is already present within the solid mix prepared for melting, less energy is required to achieve melting as there is no need to provide as much latent heat of fusion for that proportion of the solid. In order to melt crystalline materials the latent heat of fusion has to be supplied, and if glass is already present it melts with much less energy input. Secondly, the melting process is facilitated, because the initial step of melt formation from solid crystalline particles is also avoided. Once the scrap glass has melted the crystalline components of the batch can dissolve into the initial liquid with much greater ease than if they had to melt.

As already shown, the production of glass involves the liberation of carbon dioxide from the carbonate raw materials to the atmosphere. Simple sums using relative atomic masses show that in order to produce 100 tonnes of a typical glass consisting of 70% SiO_2, 20% Na_2O and 10% CaO, 122 tonnes of raw material are required, and so 22 tonnes of CO_2 are lost to the atmosphere. Consider the statistics at the beginning of this chapter: 50 million tonnes of glass are produced annually in Europe, the USA and Japan. Production of this amount of soda–lime–silica glass would yield over 10 million tonnes of CO_2. This can be reduced by diluting the raw material batch with scrap glass (cullet), which has already lost its carbon dioxide during its initial manufacture. Reduction in CO_2 emissions is a powerful argument in favour of increased use of recycled glass, in addition to the associated energy savings.

Cement and plaster | 7

The essential characteristic of cement and plaster is that they are calcined materials which harden on reaction with water when mixed as a paste. Such materials have been known and exploited since ancient times. The Egyptians used cement when building the Pyramids, and the Romans and Greeks used a cement produced from a mixture of volcanic ash and lime, which still forms the basis of some modern cement formulations (pozzolanic cements). The most widely used cement type, Portland cement, was patented in England in 1824 by Joseph Aspdin who produced it by calcining impure (argillaceous) limestones, some of which have become known subsequently as cementstones. The name Portland cement arises from the similarities in appearance of concrete made from the cement to Portland stone. Portland cement is stronger than pozzolanic cement, and has improved setting characteristics. The cementstones which were initially used to make Portland cement are widespread throughout the geological succession, and it is interesting to speculate whether they were exploited by the Romans in the northern extremities of the Roman empire, far from Mediterranean sources of volcanic ash. Plaster was also introduced to Britain by the Romans, but was lime-based; 'modern' gypsum-based Plaster of Paris was introduced in 1254.

The manufacture of cement as a clinker which is ground to give a powder as the finished product, and the processes involved in its hydration, are of major economic importance and so have been subjected to considerable study. Particularly valuable sources of further information include, despite its age, the textbook by Bogue (1955), Bye's brief but authoritative book on Portland cement (Bye, 1983), Lea's exhaustive text (Lea, 1970) and Barnes (1983).

7.1 MANUFACTURE OF PORTLAND CEMENT

Portland cement is essentially composed of three dominant chemical components, lime (CaO), silica (SiO_2) and alumina (Al_2O_3). Silica and

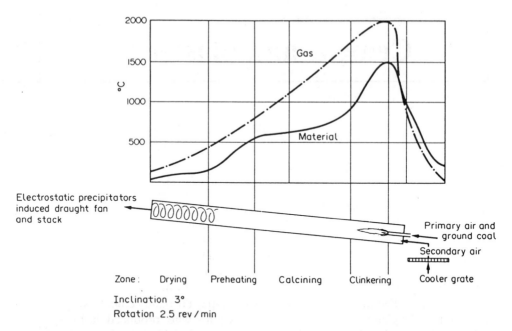

Figure 7.1 Reaction zones and temperature profiles for gas and clinker (material) in a wet-process cement kiln (from Bye, 1983).

alumina are the dominant constituents of shale, and so the raw materials for cement manufacture include essentially limestone (or chalk) and shale, which are reacted together by calcining at about 1500°C. Additional components (such as iron) can be tolerated or are required to be present for their beneficial effects in the formation of clinker compounds. Other impurities may have adverse effects on the formation of the cement or on the properties of concrete, such as magnesium. Most potentially noxious acidic gas emissions during kiln treatment are absorbed by the incoming partially calcined feed material, allowing the use of kiln fuels with a relatively high sulphur content, whether solid, liquid or gaseous.

Calcining is carried out in long, inclined, rotating kilns (Fig 7.1), through which the raw materials gradually move, with the temperature rising steadily and then cooling rapidly. There are a number of individual reactions which take place in sequence, and the rate of flow of the ingredients through the kiln needs to be controlled in order to optimize their thermal trajectory. The raw material can be introduced as a dry powder or wet, as a slurry. Dry processing is preferred because it avoids the need to remove water prior to calcining, and so requires less energy.

Heat treatment is carried out in an inclined rotary kiln which is usually preceded by some form of preheater. The process involves evaporation of any water, thermal decomposition of clay minerals (300–650°C), decomposition of calcite (>800°C) and culminates in the formation of a quantity of liquid phase and sintering at approximately 1450°C.

The clinker formed at this temperature is then cooled and removed from the kiln for grinding and finishing prior to sale. Fuel combustion (usually coal) is used to provide the heat required and coal ash may be incorporated chemically in the clinker: note that the temperature of the raw material lags behind and does not rise at the same rate as the gas temperature, reflecting in part the overall endothermic nature of the dehydration and clinker-forming reactions. Organic matter within the clays contributes to combustion, assisting the energy budget; of a notional £50 per tonne sales price, approximately 2% reflects the cost of the materials and 25% the energy costs of firing. Electricity costs for grinding of raw materials, fuel and the clinker product are rapidly becoming of equal importance to those of fossil fuel energy in several countries. In view of this alternative sources of energy, such as waste tyres or solvents, are becoming increasingly of interest to manufacturers.

Calcium oxide is a highly reactive material, and once formed it reacts with the dehydrated clay to give one or more of a number of calcium aluminium silicates. These form the cement clinker minerals, which are occasionally found in nature, especially in contact metamorphosed impure limestones.

Calcining at temperatures of approximately 1450°C allows partial melting to take place, and this facilitates reactions between the solid mineral phases which are wetted by the flux. At this temperature the lime retains the morphology of the finely-ground limestone starting material, and is at its most reactive. At higher temperatures the lime recrystallizes and coarsens to become less reactive, and is described as 'hard burnt'. All solid raw materials and fuels are very finely ground as an essential requirement for reaction.

In order to simplify long formulae, cement compounds are described in the abbreviated fashion explained in Chapter 5, and are dominated by combinations of the following oxides:

CaO = C
Al_2O_3 = A
SiO_2 = S
H_2O = H
Fe_2O_3 = F
MgO = M

As there is abundant scope for confusion between the abbreviated and conventional systems, great care has to be taken to be consistent in their use. In this chapter, reactions will be written first using the abbreviated system, and repeated using conventional notation in italics.

Table 7.1 Mineral components of cement clinker

Mineral	Abbreviated formula	Full chemical formula	Typical proportion in cement (wt %)
Tricalcium silicate (alite)	C_3S	Ca_3SiO_5	45
Dicalcium silicate (belite)	C_2S	Ca_2SiO_4	27
Tricalcium aluminate	C_3A	$Ca_3Al_2O_6$	11
Tetracalcium aluminoferrite	C_4AF	$Ca_4Al_2Fe_2O_{10}$	8

A number of minerals are characteristic of freshly made cement clinkers (Table 7.1), including the imaginatively named alite and belite. Portlandite, $Ca(OH)_2$ or CH, occurs in hydrated cements.

The chemical reactions which take place during cement manufacture can be summarized as follows:

$$2C + S = C_2S \quad (7.1)$$
$$2CaO + SiO_2 = Ca_2SiO_4$$
$$\text{lime} + \text{silica from shale} = \text{belite}$$

$$3C + A = C_3A \quad (7.2)$$
$$3CaO + Al_2O_3 = Ca_3Al_2O_6$$
$$\text{lime} + \text{alumina from shale} = \text{tricalcium aluminate}$$

$$4C + A + F = C_4AF \quad (7.3)$$
$$4CaO + Al_2O_3 + Fe_2O_3 = Ca_4Al_2Fe_2O_{10}$$
$$\text{lime} + \text{alumina} + \text{ferric oxide} = \text{tetracalcium aluminoferrite}$$

$$C + C_2S = C_3S \quad (7.4)$$
$$CaO + Ca_2SiO_4 = Ca_3SiO_5$$
$$\text{lime} + \text{belite} = \text{alite}$$

Reaction 7.4 is catalysed by the presence of C_4AF, and produces tricalcium silicate, which is responsible for the strength of concrete and is therefore an essential component. Indeed, a Portland cement would consist ideally only of calcium silicates, but it is not possible to produce a single phase economically.

Because of the simplicity of the components which make up cement, they can easily be represented on the phase diagram for the system CaO–Al_2O_3–SiO_2. A number of phases occur within this system (Figure 7.2), but

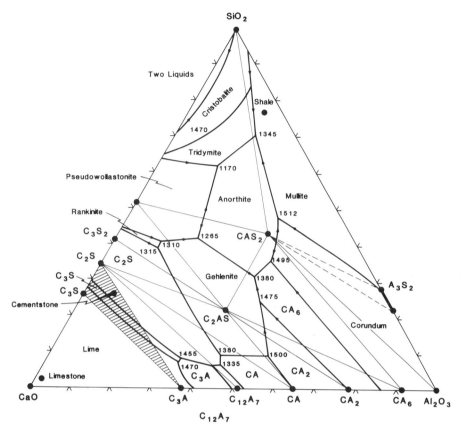

Figure 7.2 Phase relationships at one atmosphere pressure in the carbon dioxide- and water-free system C–A–S, showing plotting positions of limestone, cementstone and shale, and highlighting the C_2S–C_3S–C_3A triangle. Compositions in this diagram are expressed as weight %.

those of importance (C_3S, C_2S and C_3A) lie within a restricted part of the diagram, close to the CaO (lime) apex.

Within this diagram are plotted typical compositions for limestone (which lies close to the CaO apex) and shale (which lies close to the Al_2O_3–SiO_2 join towards the SiO_2 apex). Thus the bulk composition of the mixture of limestone and shale is chosen to lie within the C_3S–C_2S–C_3A triangle, to make sure that the required mineral phases are produced on firing. The proportions of the two rocks required to give a particular mixture can be estimated by drawing a line between the plots of their compositions and by using the lever rule (explained in Chapter 5; Figure 7.3). Note that the proportions obtained by this construction relate to lime (CaO) and anhydrous shale. Correction must be made for the loss of carbon dioxide, to deduce the proportion of limestone, as well as for the

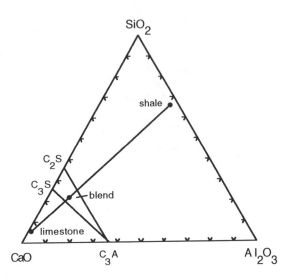

Figure 7.3 Estimation of proportions of limestone and shale required to make cement. In this example, a blend of 66% lime and 34% shale will give the required composition. If we assume that the shale volatile content is 10% and that the limestone is pure calcium carbonate (i.e. 56% CaO; 44% CO_2), the proportions of limestone and shale are corrected to 76% and 24% respectively. However, this does not compensate for the shale's lime content (see text).

loss of water (and other volatile constituents, including sulphides as sulphur dioxide) from the shale (Chapter 5; Table 5.2). A more rigorous method of assessing the proportions of limestone and shale that are required, which takes into account the lime content of the shale and the silica and alumina (and ferric oxide) contents of the limestone, is explained later in this chapter.

It is most important that cement compositions lie within the C_3S–C_2S–C_3A triangle. As this is adjacent to a three-phase triangle including CaO (instead of C_2S) care has to be taken to avoid excess CaO (i.e. to prevent oversaturation with respect to lime), because this component causes expansion of mortar when it hydrates to form portlandite. Similarly, inclusion of $C_{12}A_7$, a consequence of insufficient CaO, is to be avoided – this phase is not hydraulic (i.e. it does not react with water). The role of C_3A is beneficial for the formation of the clinker, as it is responsible for reducing the binary minimum from 1455°C for the system CaO–SiO_2 to 1335°C for the ternary eutectic in this part of the system CaO–Al_2O_3–SiO_2. Thus, on heating, a bulk composition within the C_3S–C_2S–C_3A triangle will start to melt at 1335°C, providing a liquid which will react with solids present to produce the final phase composition appropriate for the temperature of calcining.

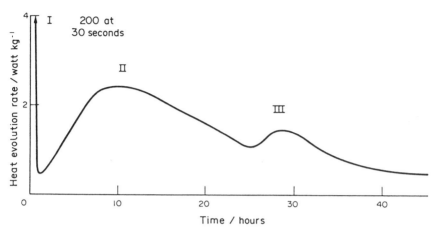

Figure 7.4 Variation in rate of heat evolution as Portland cement sets (from Bye, 1983).

Once it leaves the kiln, the cement clinker is ground and mixed with 4–7% gypsum, which helps inhibit setting while the cement is being worked. This mixture does not set for a number of hours. National quality standards may allow the grinding of proportions of other materials with the cement and gypsum if they show hydraulic or pozzolanic properties, or otherwise usefully contribute to concrete properties. Examples include natural pozzolans, granulated (quenched) blast furnace slag, pulverized fuel ash, silica fume and fine limestone, provided that suitable quality criteria are met. Some by-product gypsums can be used as set control agents, and a quantity of bassanite (see later in this chapter) can be tolerated.

7.2 SETTING OF PORTLAND CEMENT

The setting of cement is a complex process involving a number of stages which represent different mineralogical reactions which take place at different rates. It is essentially an exothermic process, and the heat evolved can be measured and correlated with the changes in physical properties of the cement. After an initial maximum there is a dormant period followed by a second maximum about 10 h after mixing, and a third at about 30 h (Figure 7.4). After about 40 h heat evolution cannot be detected. The second maximum correlates well with increasing stiffness of the mix – the point at which the cement sets.

A complex series of reactions takes place as the cement reacts with water:

7.2.1 C_3S

Setting of C_3S involves hydration and the formation of portlandite and a calcium silicate hydrate:

$$C_3S + (y+z)H = zCH + C_{(3-z)}SH_y \qquad (7.5)$$

$$Ca_3SiO_5 + (y+z)H_2O = zCa(OH)_2 + Ca_{(3-z)}SiO_{(5-z)} \cdot yH_2O$$

The calcium silicate hydrate is poorly crystalline. It has a very high surface area, and can be regarded as a gel. If this compound is hydrated in isolation it shows a single exothermic maximum at about 10 h corresponding to the second maximum when cement sets. The presence of the dormant period before C_3S reaction has caused much discussion, and may be due to several factors. Of these, perhaps the most important is the fact that the addition of water to cement yields a solution saturated in lime, and this needs to precipitate as CH before C_3S can react.

Electron microscopy of setting C_3S shows the presence of a number of intriguing surface coatings, as tubes and gels. These may also play a role, inhibiting reactions by armouring the surfaces of grains.

7.2.2 C_2S

On hydration, C_2S shows similar behaviour to C_3S, but is slower to react. It does however continue to hydrate late in the setting period, and may then contribute to the strength of the cement.

7.2.3 C_3A

Compared with tricalcium silicate, tricalcium aluminate reacts very rapidly with water, giving two crystalline hydrated products:

$$2C_3A + 21H = C_4AH_{13} + C_2AH_8 \qquad (7.6)$$

$$2Ca_3Al_2O_6 + 21H_2O = Ca_4Al_2O_7 \cdot 13H_2O + Ca_2Al_2O_5 \cdot 8H_2O$$

These form hexagonal platelets within the cement, and convert at above 30°C to the cubic hydrogarnet C_3AH_6, which forms very quickly, and is responsible for the initial formation of a crystalline network:

$$C_4AH_{13} + C_2AH_8 = 2C_3AH_6 + 9H \qquad (7.7)$$

$$Ca_4Al_2O_7 \cdot 13H_2O + Ca_2Al_2O_5 \cdot 8H_2O = 2Ca_3Al_2O_6 \cdot 6H_2O + 9H_2O$$

Some of this may be formed immediately under normal conditions if the initial exothermic reactions bring the temperature up to 30°C. In the presence of excess lime in solution, as is normal in cements, the formation of C_4AH_{13} is favoured. This slows the formation of C_3AH_6, but even so the formation of C_4AH_{13} can be a problem, causing the cement to set too quickly. Consequently gypsum is added, and the mineral ettringite ($C_6A\bar{S}_3 \cdot 31-32H_2O$ (where $\bar{S}$ denotes sulphur trioxide, SO_3) or $Ca_6Al_2O_6(SO_4)_3 \cdot 31-32H_2O$) is formed on hydration:

$$C_3A + 3C\bar{S}H_2 + 25-26H = C_6A\bar{S}_3H_{31-32} \quad (7.8)$$
$$Ca_3Al_2O_6 + 3CaSO_4 \cdot 2H_2O + 25-26H_2O = Ca_6Al_2O_6(SO_4)_3 \cdot 31-32H_2O$$

tricalcium + gypsum + water = ettringite
aluminate

Ettringite resembles the silicate zeolites (Chapter 4) in that it has a very open acicular structure made up of columns of Ca and Al with water in channels between the columns.

Again, in detail, electron microscopy reveals the formation of colloidal calcium aluminates and silicate gels which coat individual grains and increase in thickness during the initial making of the cement and mixing of concrete or mortar, allowing the bulk mixture to form a paste. As the cement ages, these colloidal coatings spall off and the cement recrystallizes to form a relatively weak three-dimensional network of $Ca(OH)_2$ and C_3AH_6: it becomes solid, set and no longer workable. The cement then develops its strength as the crystals coarsen and form an interlocking network.

7.2.4 C_4AF

This phase is more strictly represented by the formula $xC_2A \cdot (1-x)C_2F$. It behaves similarly to C_3A initially to form complex hydrated products:

$$C_4AF + 13H = C_4AFH_{13} \quad (7.9)$$

where C_4AFH_{13} is a hexagonal, low temperature phase. If excess lime is present it is followed by the formation of cubic hydrogarnets which show solid solution between C_3AH_6 and C_3FH_6. In the presence of gypsum, an Fe-bearing ettringite is formed.

Overall, setting involves a combination of all of these reactions. The critical aspects include the formation of a C–S–H phase and the cubic hydrogarnets. It is the interlocking nature of the C–S–H phases in particular which provides the strength of the cement. Note that reaction does not go to completion, and that some of the clinker minerals may persist as relic phases in the cement.

7.3 SPECIAL CEMENTS

Portland cement provides a 'standard' cement for construction work. Special cements are produced for particular requirements. Some examples are given below:

7.3.1 Quick setting cement

Portland cement ground with a clinker rich in $C_{12}A_7$ or the fluorinated equivalent $C_{11}A_7 \cdot CaF_2$ sets within 30–60 minutes, but achieves its full strength only after 28 days. Clinkers enriched in this way are achieved by the addition of bauxite ($Al_2O_3 \cdot 2H_2O$) and/or fluorite to the raw material mixture introduced to the calciner.

7.3.2 High alumina cement

This silica-poor cement is made by total melting of iron-bearing bauxite and limestone at about 1700°C in an open furnace. The resulting phases are dominated by CA, CA_2, C_2AS and C_2S. Under ideal conditions, concrete made with high alumina cement develops full strength much more rapidly than if Portland cement had been used. Its setting involves the hydration of CA to give CAH_{10}, which then reacts further to give an aluminium hydroxide gel which forms gibbsite (AH_3 or $Al(OH)_3$):

$$CA + 10H = CAH_{10} \qquad (7.10)$$
$$CaAl_2O_4 + 10H_2O = CaAl_2O_4 \cdot 10H_2O$$

$$3CAH_{10} = C_3AH_6 + 2AH_3 + 18H \qquad (7.11)$$
$$3CaAl_2O_4 \cdot 10H_2O = Ca_3Al_2O_6 \cdot 6H_2O + 4Al(OH)_3 + 18H_2O$$

The recrystallization of CAH_{10} involves a loss in volume of the solids, increasing concrete porosity and reducing strength. High alumina cement is generally no longer used in load-bearing applications.

7.3.3 Sulphate-resistant cement

This is a Portland cement designed for use in situations where the groundwater has a high sulphate content, where soils contain sulphates such as gypsum, or where sea water may come into contact with the cement. Sulphate-resistant cement has a C_3A content below 3.5%, achieved by lowering the alumina content of the raw material mix to 70%. This inhibits the formation of C_3A, which reacts with sulphate to produce ettringite causing swelling and failure as a consequence of the associated expansive forces.

7.3.4 Low-alkali cements

Cements with a low alkali content may be required for use in the manufacture of concrete in which the use of a particular aggregate (e.g. sea-dredged gravel) introduces alkalis, giving grounds for suspicion that alkali–silica reactivity might occur. Low alkali cement is designed to have an equivalent maximum Na_2O content of 0.6% (calculated as wt% Na_2O + 62/94 K_2O).

7.4 SELECTION AND BLENDING OF RAW MATERIALS

Geologically, the constraints on raw materials are relatively simple. Limestone and shale are the major raw materials, and their proportions can be determined, to a first approximation, using the phase diagram for the system $CaO-Al_2O_3-SiO_2$, provided that their compositions are known (as shown in Figure 7.3). The blended mixture of the two rocks must plot within the $C_3S-C_2S-C_3A$ triangle. However, this approach neglects the iron content of the rocks, and so an equivalent exercise ought to be carried out using the quaternary system $CaO-Al_2O_3-Fe_2O_3-SiO_2$. This is not practical as a graphical exercise, and instead the calculated lime saturation factor is used to constrain the blend. For this, it is essential to know the chemical composition of the raw materials, and this is routinely determined using X-ray fluorescence analysis of borehole samples taken ahead of mining. This is particularly important for interbedded limestones and shales, where the relative proportions, thicknesses and compositions of each rock type must be known not only to permit correct blending but also in determining the choice of sample size used to assess the deposit. X-ray fluorescence methods are also used to monitor routinely the composition of the finished cement product, in part to allow compensation (where necessary) for the alkali content of aggregate to be mixed with the cement to form concrete.

The **lime saturation factor** (LSF) is calculated from the chemical analysis of the raw materials as follows:

$$\text{LSF} = \frac{CaO}{2.80\ SiO_2 + 1.18\ Al_2O_3 + 0.65\ Fe_2O_3} \quad (7.12)$$

where the oxide formula denotes the weight % content of that oxide.

The LSF is a measure of the ability of the blend to react leaving no free lime; i.e. the solid phases are saturated (overall) with respect to CaO. When LSF equals 1, the amount of lime exactly balances the amount of silica, alumina and ferric oxide and so $CaO = (2.80 \times SiO_2) + (1.18 \times Al_2O_3) + (0.65 \times Fe_2O_3)$: this formula describes the plane in the quaternary system C–A–S–F which corresponds to 100% lime saturation. If the lime saturation factor is in excess of 1, free lime will be present within the produced clinker. In practice, materials are blended to give a value for the LSF typically of 0.96 (or 96%). The blending of raw materials takes into account the amount of silica, alumina and ferric oxide in the limestone

Table 7.2 Typical compositions (wt %) of limestone and shale used for cement manufacture, with the composition of a blend produced according to the requirement to give a lime saturation factor of 96% (see text for details) and proportions (%) of the ternary components required for plotting these data in Figure 7.3

Component	Limestone	Shale	Blend
SiO_2	3.00	50.10	13.41
Al_2O_3	0.70	22.90	5.61
Fe_2O_3	0.30	7.90	1.98
CaO	53.70	2.80	42.45
MgO	0.20	2.50	0.71
K_2O	0.10	2.40	0.61
Na_2O	0.10	0.70	0.23
SO_3	0.10	0.24	0.13
CO_2	41.60	0.00	32.41
LOI*	0.00	10.00	2.21
Totals	100.00	100.00	100.00
CaO	93.55	3.69	69.06
Al_2O_3	1.22	30.21	9.12
SiO_2	5.23	66.09	21.82
Proportions in blend	77.9	22.1	

* excluding CO_2

and the amount of lime in the shale, and is best illustrated by a worked example.

Table 7.2 gives typical compositions for a limestone and a shale. The limestone contains silica, alumina and ferric oxide which will react with a proportion of its lime content. This proportion is given by the sum of:

$(2.80 \times SiO_2) + (0.96 \times Al_2O_3) + (0.65 \times Fe_2O_3)$, which in this case is: $8.4 + 0.67 + 0.20 = 9.27$.

For a lime saturation factor of 0.96, the amount of lime required to saturate these oxides is $0.96 \times 9.27 = 8.9$ equivalent parts. So, the lime content of the limestone which is available to react with the shale becomes $53.7 - 8.9 = 44.8$ equivalent parts.

Similarly, the shale content of silica, alumina and ferric oxide can be calculated as:

$(2.80 \times SiO_2) + (0.96 \times Al_2O_3) + (0.65 \times Fe_2O_3)$, giving: $140.3 + 21.98 + 5.14 = 167.4$.

Again this is corrected by a factor of 0.96, to become 160.7 equivalent parts. As this shale contains 2.8% lime, the available silica, alumina and ferric oxide which can react with lime from the limestone becomes 160.7 − 2.8 = 157.9 equivalent parts.

Thus, in this example, for complete reaction, one unit of shale requires 157.9 equivalent parts of lime, but one unit of limestone can only provide 44.8 equivalent parts of lime. Thus the amount of limestone required to satisfy one unit of shale is given by the ratio limestone/shale = 157.9/44.8 (3.525:1 or 77.9% limestone and 22.1% shale).

Further corrections can be made for the addition of other raw materials into the blend, such as loam (to increase the silica content) or coal ash (derived from coal used as a kiln fuel; Bye, 1983). These corrections may seem to be tiresome, but they are essential to ensure that the finished cement consists of the correct assemblage of cement minerals.

There are in addition a number of other constraints on the composition of the raw material blend required for the manufacture of Portland cement. These concern compensation for the following impurities:

- Mg   from dolomite; if in excess of 1.5% MgO, periclase is formed, which hydrates after setting and expands
- F   from fluorite; setting is retarded if F>0.2%
- alkalis   may enhance reactions with amorphous silica in aggregates
- sulphur compounds   form calcium sulphate, reducing the need to add gypsum

Cement factories are widely distributed throughout the world, but tend to be located on the basis of geological factors which permit the mining of adjacent (or interbedded) deposits of limestone (or chalk) and shales. This reduces the need to pay transport costs for imported bulk raw materials in addition to the costs of transporting the product to the markets. Britain is typical in this respect. Many of the 19 British cement works which were active in 1991 were situated close to geological boundaries between limestones and shales, so that both could be mined within a single property, albeit from different pits in some cases (Figure 7.5). A wide range of geological ages are involved, including the Devonian, Carboniferous, Jurassic and Cretaceous.

Because blending has to be carried out so carefully for successful cement manufacture, it is essential to understand the spatial variation of the chemical composition of the raw material. Figure 7.6 illustrates the way in which limestone quality can be described by dividing a deposit into blocks, and emphasizes the way in which folding (or other geological structures) influences the distribution of grade relative to the requirements of quarrying for benching. We will return in Chapter 10 to consideration of how to evaluate variable deposits of this type, but Figure 7.6 emphasizes the need to know the extent of spatial variation in raw material quality. In

CEMENT AND PLASTER

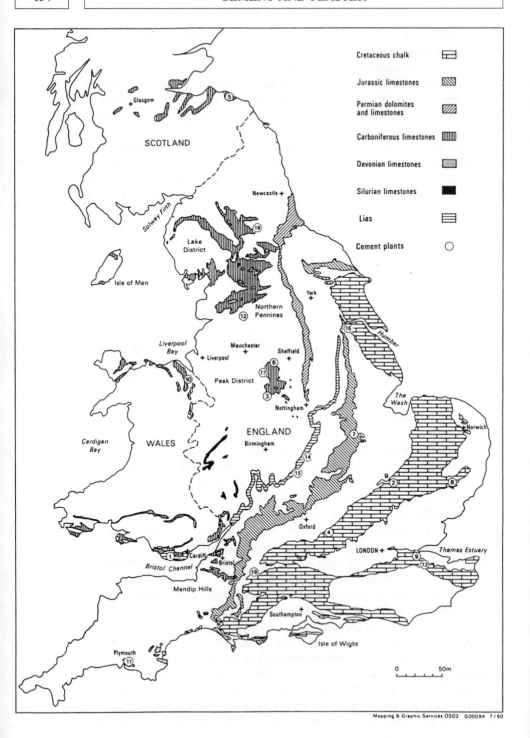

addition to the raw materials quarried by the cement manufacturer, various by-products and wastes may also be incorporated from other sources, provided the chemistry is appropriate. It is becoming increasingly frequent for manufacturers to use minor additions of high purity limestone, iron ore (or waste oxide), sand or bauxite to correct for deficiencies in the quarried materials.

7.5 PLASTER MINERALOGY AND PRODUCTION

Plaster was introduced to Britain by the Romans, but was lime-based; gypsum-based plaster was introduced in 1254 and was known as Plaster of Paris as the Montmartre quarries were the best known source. It is produced by calcining gypsum at 150–165°C:

$$CaSO_4 \cdot 2H_2O \rightarrow CaSO_4 \cdot 0.5H_2O \qquad (7.13)$$

This hemihydrate is known as the mineral bassanite, and rehydrates to form gypsum once water is added.

Calcined gypsum is used to produce a number of different plaster products:

- pre-mixed lightweight plaster – bassanite with a lightweight aggregate such as vermiculite or perlite.
- retarded hemihydrate plaster – used as a finishing plaster, or with the addition of sand as a heavy undercoat plaster. Chemical retarders are used to slow down setting times.
- plasterboard – hemihydrate plaster made into sandwiches with paper or board facings; a slurry is produced, sandwiched by paper and then dried. Mineralogically, bassanite has reverted to gypsum and plasterboard is effectively a gypsum sandwich.
- 'baritite' – the addition of baryte (barium sulphate) aggregate is used to produce plaster suitable for use in radiological protection, for example to plaster rooms used for X-ray equipment.

There are few constraints on the purity of gypsum required for plaster products. Occuring as an evaporite mineral (Chapter 4), gypsum is often

---

Figure 7.5 Location of cement factories (1991) relative to outcrops of limestone, dolomite and chalk in Britain. 1. Aberthaw (South Glamorgan); 2. Barrington (Cambridgeshire); 3. Cauldon (Staffordshire); 4. Chinnor (Oxfordshire); 5. Dunbar (Lothian); 6. Hope (Derbyshire); 7. Ketton (Leicestershire); 8. Masons (Suffolk); 9. Northfleet (Kent); 10. Padeswood (Clwyd); 11. Plymstock (Devon); 12. Ribblesdale (Lancashire); 13. Rochester (Kent); 14. Rugby (Warwickshire); 15. Southam (Warwickshire); 16. South Ferriby (Humberside); 17. Tunstead (Derbyshire); 18. Weardale (Durham); 19. Westbury (Wiltshire). (From Department of the Environment, 1991a.)

CEMENT AND PLASTER

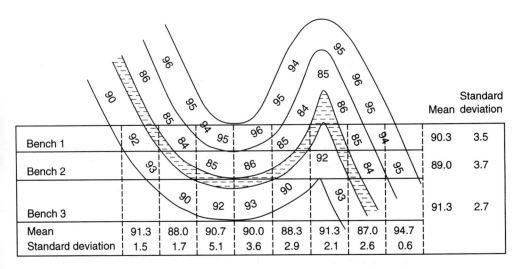

Figure 7.6 Spatial variation in grade as a consequence of folding. The limestone beds are divided into blocks with a particular calcium carbonate content, and the mean and standard deviation figures refer to the quality of the material mined from the benches. In this example, when benches and bedding coincide standard deviations are small, but when folding results in beds of different quality occurring on a given bench standard deviations increase, demonstrating greater variability in quality of mined output (courtesy of Blue Circle plc).

associated with red-bed sediments and may be contaminated by haematitic clays. In Britain, gypsum from these sources yields a plaster which is pink in appearance, due to the persistence of finely disseminated haematite.

Gypsum is produced widely throughout the world, largely to satisfy domestic markets; demand is greatest in the United States (producing 18 million tonnes and consuming 27 million tonnes in 1990). The leading exporters are France, Spain, Canada and Thailand, where production exceeds domestic demand. Gypsum deposits of a wide variety of geological ages are mined, from Lower Palaeozoic and Carboniferous rocks in North

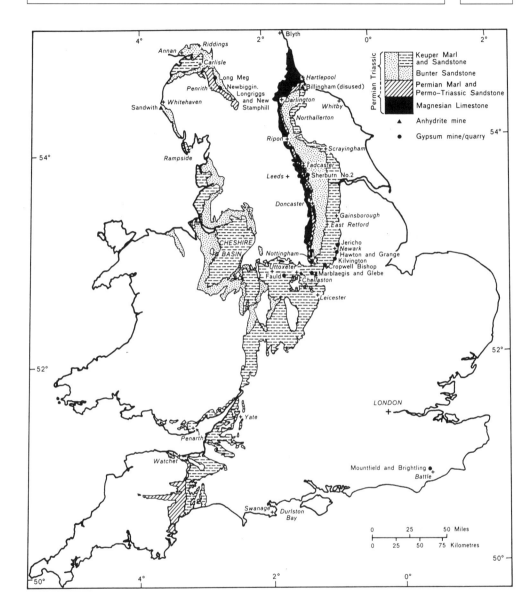

Figure 7.7 Occurrence of gypsum-bearing strata and mining areas in Britain (from Notholt and Highley, 1975).

America, the Permo-Trias in Britain and Germany, and the Tertiary of the Paris Basin and Rhine Graben.

Geologically, the occurrence of gypsum of industrial interest in Britain is concentrated in the Permian sequences of the Vale of Eden and south

Nottinghamshire (Figure 7.7). In the Vale of Eden gypsum (and anhydrite) occur within upper Permian evaporites between the Penrith and St Bees sandstones, in the Eden Shales. Individual units are up to 36 m thick, and are mined underground or opencast. In north-east England, Zechstein evaporites were mined for anhydrite at Billingham. In the East Midlands, gypsum occurs throughout the Mercia Mudstone Group (Keuper Marl) of Triassic age. They are mined at a number of sites, underground and at surface, near the top of the Keuper sequence. Individual seams reach 4.5 m in thickness (Tutbury Gypsum) and up to about 1 m within a 15–18 m thick gypsum-rich zone known as the Newark Gypsum. The occurrence of gypsum is restricted to near surface, because at depth it passes into anhydrite. Its formation in the British Triassic is in part a consequence of anhydrite hydration due to the interaction with meteoric waters.

In addition to mined gypsum, an alternative source of gypsum suitable for plaster manufacture is provided by flue gas desulphurization plant installed in coal-burning electricity generating stations. The quantities involved are large: a single power plant may yield 250 000–1 million tonnes of gypsum annually, capable of supporting a plasterboard factory and making up a significant component of gypsum consumption.

Clays for construction   8

In the construction industry, clays are a vital raw material for the production of bricks, tiles and earthernware pipes. For thousands of years bricks have been made out of mud, clay and shale which have been chosen largely on the basis of the criterion that 'if it makes a brick, it's a brick clay', and there has been a tendency, historically, to make do with the product, as long as it is strong enough for the job. However, in the eyes of the sophisticated consumer in a modern industrial society, a brick must be aesthetically pleasing as well as meeting in-service strength and durability requirements, and many purchasing decisions depend on the selection made by the architect. A successful brick company will therefore aim to produce a wide range of products which vary in colour and surface texture, perhaps from a very limited selection of clay raw materials. Price is important; at the time of writing (1993) a target for a typical brick range would be to be able to sell at a profit 1000 bricks for £100–£120 or less. It is important for the geologist to realize that it is success in the sale of particular ranges of brick that ultimately dictates demand for particular clay raw materials, and that this provides an important incentive in the search for satisfactory alternatives to existing sources which might be drawing near to exhaustion. Although this chapter concentrates on brick manufacture, the same principles apply to the chemical and mineralogical changes involved in the formation of earthenware pipes and tiles (roof and floor tiles), which differ principally in the way in which they are formed and fired.

Historically, bricks have been produced from a large number of local brickyards situated close to towns and villages, each producing often distinctive bricks from local clays. There may no longer be any evidence of these brick works, other than allusions to their former presence in the names of roads and topological features, but their widespread distribution in Britain at the beginning of the twentieth century is indicated by the distribution of clay workings shown in Figure 8.1. As industrialization has proceeded, brick production has been rationalized to fewer centres of

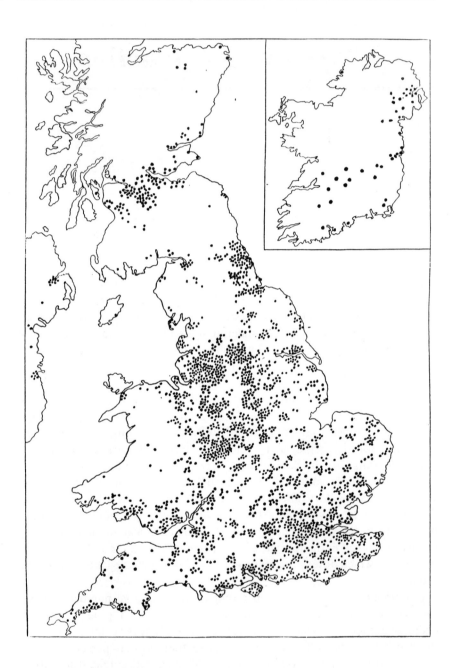

Figure 8.1 Distribution of 'clayworks' in Britain and Ireland in the early 1990s (from Searle, 1912). Note that this shows all clay workings, including the china clay and ball clay works of south-west England.

production (Figures 8.2 and 8.3), and this process continues with the current recession in the construction industry in Britain (Chapter 1). A consequence of centralization of production has been an increase in regional trade in shale and clay, as clays may need to be brought in to provide suitable material for blending to achieve production of a range of brick products which vary in their appearance.

8.1 RAW MATERIALS

Clays and shales suitable for brick manufacture are very widespread on the Earth's surface. They contain clay minerals as essential constituents, together with other minerals and rock fragments. The nature of a brick depends in detail on the reactions which take place when the inhomogenous bulk clay raw material is fired.

The important clay minerals kaolinite, illite and smectite have been introduced in Chapter 3, in a discussion of the relatively pure industrial clays. All three can occur as components of brick clays, together with members of the chlorite group, such as chlorite ($Mg_{12}Si_8O_{20}(OH)_{16}$) and clinochlore ($Mg_{10}Al_2Si_6Al_2O_{20}(OH)_{16}$), which are magnesium-bearing clays (within which Fe can substitute for Mg).

In addition to clay minerals, brick clays may contain other mineral constituents.

- *Fe–Al oxides and hydroxides* – Fe and Al form structurally analogous oxides and hydroxides, which are derived from weathered or altered ferromagnesian and aluminosilicate minerals (n.b. in weathering, magnesium enters solution rather than forming a solid oxide or hydroxide phase):

gibbsite	$Al(OH)_3$		
diaspore	α-AlO·OH	goethite	α-FeO·OH
boehmite	γ-AlO·OH	lepidocrocite	γ-FeO·OH
		haematite	Fe_2O_3
		limonite	$2Fe_2O_3\cdot3H_2O$

- *Sulphides* – pyrite (FeS_2) is a common constituent of black shales, either in a finely disseminated form or within concretionary nodules.
- *Carbonates* – calcite ($CaCO_3$), and siderite ($FeCO_3$) also both occur as concretions, and calcite is present as fossil (or recent) shell material, lumps of limestone in till (boulder clay), diagenetic cement in sandstone fragments or as disseminated material. Dolomite ($CaMg(CO_3)_2$) may also occur.
- *Sulphates* – gypsum ($CaSO_4\cdot2H_2O$) is potentially an important constituent of clays quarried from evaporitic sequences (see Chapter 7 for discussion of plaster).

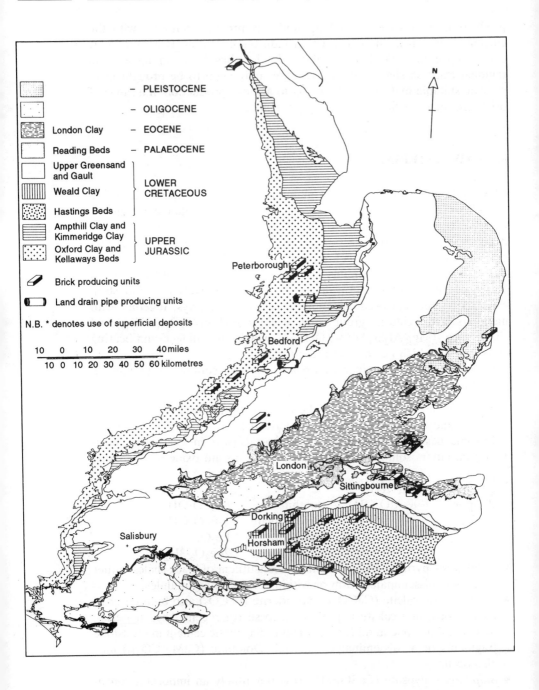

Figure 8.2 Distribution of brick and tile works and the outcrop of major clay-producing sequences in south-east England (Ridgeway, 1982).

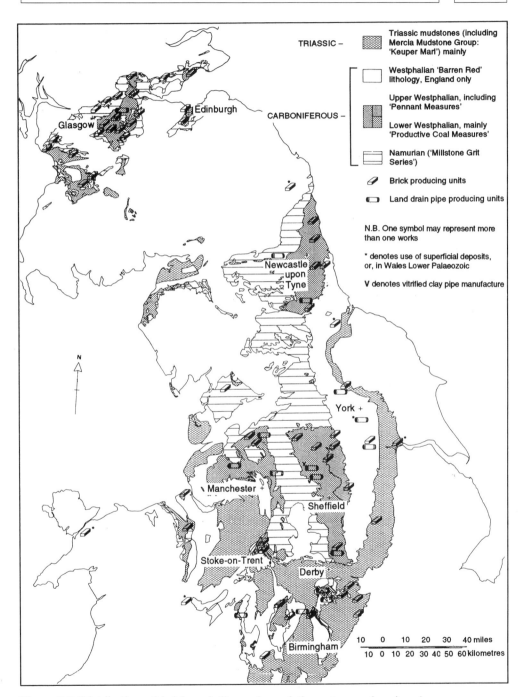

Figure 8.3 Distribution of brick and tile works and the outcrop of major clay-producing sequences in northern England and Scotland (Ridgeway, 1982).

- *Residual minerals* – quartz (SiO_2) is the most common residual mineral. Most aluminosilicates and ferromagnesian silicates are unstable during weathering and decay to form clays. Minor traces of these which have survived weathering may be present.
- Organic carbons – organic matter (hydrocarbons or coal) is commonly present within brick clays.

Each of these components can play a part in brick firing, which essentially involves reactions in which the clay minerals break down, sinter and fuse to form a 'cement' which binds other minerals and rock fragments together. Although the strength and appearance of the brick may depend, at least in part, on the presence of minerals other than clays, it is the behaviour of the clay mineral assemblage which is important for the formation of the brick.

The thermal decomposition of clays is critical in the process of firing, and different clay minerals individually vary in their behaviour. In addition, the proportion and nature of the clay mineral assemblage varies from one bulk clay to another, depending on the age and depositional history of the rock of interest (Figure 8.4). This variability gives rise to variation in the firing behaviour of different raw materials.

Even though the nature of the clay fraction is so important in predicting the firing behaviour of a bulk clay, it is rarely known in detail as the cost of undertaking a detailed mineralogical analysis of the clay fraction cannot necessarily be justified, given the need to charge a low price for the finished product. Instead, brick clays are characterized by bulk chemical analysis, especially X-ray fluorescence (which is sufficiently cheap that heterogeneity in raw material supply can be assessed routinely). Thus in the exploitation and evaluation of brick clay raw materials it is not always practical to have a detailed knowledge of the clay mineralogy, and it is necessary instead to interpret a bulk chemical composition using the assumptions that kaolinite is an alkali-free clay, illite may be the dominant host for potassium (in the absence of feldspar), montmorillonite is complex but Ca- and Na-bearing, and chlorite is ferromagnesian.

In terms of their bulk chemical composition, brick clays are dominated by silica and alumina, with very variable amounts of iron, magnesium, alkalis and calcium oxide. Representative analyses are given in Table 8.1. Some elements may be present in more than one mineral, especially lime: as calcite, gypsum and/or clay (smectite). Because of possible deleterious consequences, care must be taken to determine whether the following minerals are present:

- calcite/limestone – affects the colour of the brick. If present in large fragments, there may be problems with formation on firing of CaO, which hydrates on cooling/wetting, expanding and popping the face of the brick.

RAW MATERIALS

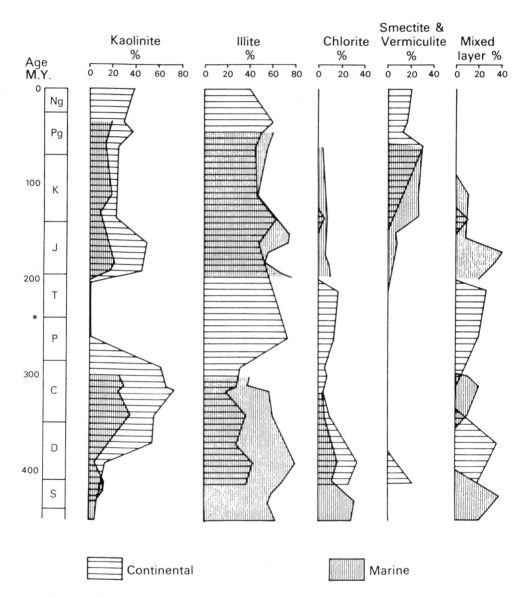

Figure 8.4 Dominant clay mineral assemblages in selected British brick clays (adapted from Ridgeway, 1982). The term 'mixed layer' refers to intergrown illite and smectite.

- gypsum — dehydrates to form bassanite, and then rehydrates on exposure to form gypsum, expanding in the process and causing a skin to form on the brick surface.
- pyrite — oxidizes on firing to produce sulphate. This reacts with lime to give gypsum, and if only mild, causes efflorescence (a superficial white encrustation of soluble salts, mostly sulphates) on brick surfaces. More seriously, the generation of sulphur dioxide gas derived from pyrite causes 'bloating' (swelling) of the brick, and contributes to flue gas emissions.

It is possible to recalculate the chemical analysis of a bulk clay to yield a mineralogical composition analogous to a CIPW norm, but this can only be done with detailed knowledge of the composition of individual minerals.

8.2 MINERALOGICAL CHANGES DURING FIRING

The process of firing involves heating a damp starting material, the bulk clay, up to temperatures where it begins to melt. Bricks are essentially

Table 8.1 Compositions of selected brick clay raw materials (wt %)

Component	1 Shale	2 Fireclay	3 Etruria Marl	4 Etruria Marl	5 Mercia Mudstone	6 Mercia Mudstone	7 Oxford Clay	8 Boulder Clay
SiO_2	53.69	56.30	58.12	51.39	48.70	65.30	45.82	63.41
TiO_2	0.20	1.16	1.35	1.27	0.69	0.74	0.84	0.70
Al_2O_3	20.50	25.50	22.40	23.10	13.30	14.00	15.21	13.81
Fe_2O_3	6.95	3.19	6.79	9.02	5.05	4.55	2.81	4.57
FeO	0.86	–	–	–	–	–	–	–
CaO	0.30	0.30	0.40	2.00	5.59	1.69	10.10	2.40
MgO	2.41	0.74	1.28	0.86	8.37	2.73	2.21	3.27
Na_2O	0.62	0.11	0.14	0.10	0.09	0.53	0.86	–
K_2O	2.73	2.13	1.65	1.79	5.07	5.03	2.62	2.34
SO_3	0.37	–	–	–	–	–	–	–
LOI	11.14	9.76	7.40	9.09	11.80	4.48	15.00	8.70
Totals	99.77	99.19	99.53	98.62	98.66	99.05	95.47	99.21

1, Weeton Shales, Namurian, Lancashire/Yorkshire (includes 0.13% FeS_2; Ridgeway, 1982); 2, Etruria Marl, Staffordshire (Ridgeway, 1982); 3, Calcareous Etruria Marl, Staffordshire (Ridgeway, 1982); 4, Fireclay, Upper Carboniferous, Leicestershire; 5, Mercia Mudstone (Keuper Marl), Triassic, Leicestershire (Ridgeway, 1982); 6, Mercia Mudstone (Keuper Marl), Triassic, Leicestershire (Ridgeway, 1982); 7, Lower Oxford Clay, Jurassic, Fletton, Northamptonshire (includes 1.83% FeS_2; Ridgeway, 1982); 8, Quaternary Boulder Clay, Cheshire.

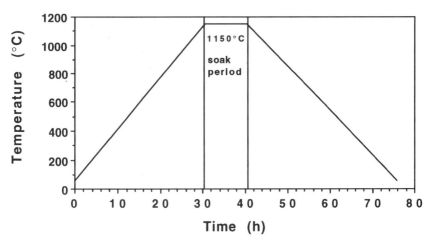

Figure 8.5 An example of an ideal temperature profile for a tunnel kiln, showing the firing cycle of approximately 3 days (76 h) and a soak period of 10 h at 1150°C.

partially molten solids which have been quenched. In geological terms, this is equivalent to taking sediments all the way through to anatectic conditions, where melting begins. Care must be taken not to fuse too much, otherwise the brick melts and loses its shape; bricks deformed by excessive heating can sometimes be found in reject piles near brickworks, or amongst stocks of reclaimed bricks dating back to Victorian and other times when quality was not monitored as closely as it is today.

Brick firing involves the use of continuous or batch kilns in which the cycle of heating and cooling takes around 2–3 days for a continuous tunnel kiln, or a number of weeks for a batch kiln (or clamp), with a 'soak' period at the firing temperature. The conditions of firing are selected on the basis of experience, and are assessed on the basis of the shrinkage of the brick or proving (buller's) rings. An ideal firing curve for a tunnel kiln is shown in Figure 8.5. In a tunnel kiln, the bricks are carried through on a continuous series of kiln cars; when one laden with unfired bricks is pushed in at one end, a car laden with finished bricks emerges simultaneously at the other end (Figure 8.6).

The manufacture of bricks involves many stages prior to firing (Figure 8.6). First, the clay is quarried and stockpiled, sometimes with storage times of several months. At this stage initial blending and mixing of clays from different sources is carried out, and the weathering of the clay which takes place may beneficially and partially remove sulphides by oxidation. The bulk clay is drawn from the stockpile, blended with other clays and additives and the required amount of water, to produce a paste that can be extruded through a die. This ribbon of clay undergoes surface roughening or dusting with coke breeze or sand (if a non-smooth surface is required),

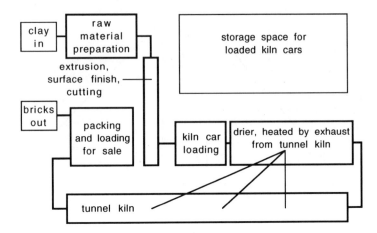

Figure 8.6 Schematic flow diagram of a brick manufacturing plant to show preparation, drying, and firing in a tunnel kiln. The kiln cars run on a railway system, with perpendicular tracks and axles to allow movement at right angles from one area to another.

and is cut up into bricks of the right shape and size. The bricks then proceed to driers, which exploit exhaust heat from the kiln, before being stacked on kiln cars prior to entry into the tunnel kiln, or they may be stacked on kiln cars prior to drying. Extruded bricks of this type usually have a pattern of holes through their centres, which reduces stress defects arising from the extrusion process as well as yielding savings in the amount of material used. Automation of most of the processes involved means that labour requirements can be very low.

Moulded bricks are prepared for the kiln in a less continuous process, as the clay has to be taken from the blending system, cast in a mould then removed prior to stacking. However, if the moulding process is carried out by hand, the resulting hand-thrown bricks can be sold at a premium price. The production of hand-thrown bricks can be reduced to involving staff at a bench taking clay from one machine and throwing it into a mould which goes to another machine. This routine job is gradually being taken over by machines capable of producing imitation 'hand-thrown' bricks.

In general, firing involves a number of stages, at the following approximate temperatures:

1. drying: up to 200°C; this involves most shrinkage
2. removal of mineral-bound water; decomposition of gypsum: 150–650°C
3. burning of carbonaceous matter in air: 200–900°C
4. decomposition of bassanite, sulphides and carbonates: 400–950°C
5. sintering and vitrification: 900°C plus

Most firing is undertaken at 1000–1100°C, with firing periods of no more than 48 h, depending on demand. If there are large stockpiles of finished bricks, the rate at which kiln cars pass through the kiln can be reduced, and the bricks spend longer being fired. Methane (natural or landfill gas, or a combination of the two) is widely used for firing in tunnel kilns, and oil or solid fuel in some older batch methods.

Cooling is carried out at a controlled rate, to avoid mechanical problems associated with the α–β quartz transition at 573°C, as the two polymorphs of quartz have differing coefficients of thermal expansion (Figure 8.7).

The firing curve shown in Figure 8.8 is given as an example observed in a study of the changes which take place during firing of a batch of bricks produced from a blend of fireclay and shale, in which partly fired bricks were removed from the kiln at access points along its length. Figure 8.8a shows that the firing curve used in this case does not have a clearly defined soak period, and that it briefly reaches a maximum temperature of approximately 1065°C. As the bricks pass through the kiln they shrink, but shrinkage increases rapidly once the temperature reaches approximately 850°C. The shrinkage curve in this case shows a number of spikes, which may not be real in view of the fact that this batch of bricks has a rough surface finish, increasing errors in the measurement of length. Rather more

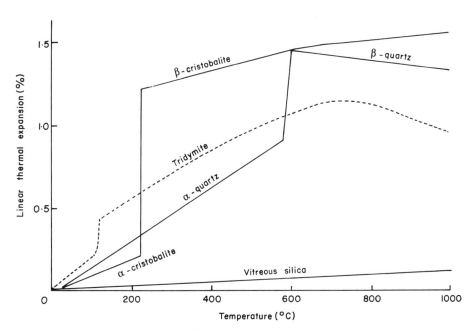

Figure 8.7 Coefficients of thermal expansion for the polymorphs of silica (from Highley, 1977).

information can be gained from the ignition losses, which indicate the amount of volatile material, mostly water, which remains to be lost as firing continues (Figure 8.8b). This figure shows that although there are spikes, there are also two plateaux where ignition losses remain more or less constant. These observations permit the firing sequence in this case to be divided into three stages: stage (1) involves steadily decreasing ignition losses (i.e. progressive loss of volatiles from the brick during firing) until reaching a temperature of approximately 850°C; stage (2) represents a plateau with little change in ignition loss which remains at approximately 4%, until the temperature reaches about 950°C, and stage (3) represents complete ignition with a residual ignition loss of no more than 0.5%. Stage (2) also marks the start of rapidly increasing shrinkage (Figure 8.8c). The transition from stage (2) to stage (3) may relate to the final decomposition of hydrous mineral phases and the formation of an anhydrous melt. However, the bulk material used in this example contains a complex mineral assemblage, reducing the level of confidence in any general interpretation of the mineralogical changes represented by the curves shown in Figure 8.8, which are specific to a particular raw material blend and brick product.

In detail, the firing behaviour of individual mineral components of the brick clay varies. Coarse rock fragments and some minerals (such as quartz) survive the firing process relatively unchanged whereas others, especially the clays, react rapidly. In any case, the process of firing does not necessarily achieve a mineral assemblage which is at thermodynamic equilibrium, and this must be remembered in attempts to predict firing properties. The clay minerals react rapidly, partly because of their fine grain-size and large surface area, and partly because of their inherent instability at high temperatures. The firing behaviour of individual clay mineral species is however poorly known, with few published data except for kaolinite (Figure 8.9). Kaolinite undergoes a number of transformations. Between 400°C and 700°C it loses water to become metakaolinite (an anhydrous phase, $Al_2Si_2O_7$), which then transforms at around 900–1000°C to produce mullite ($Al_6Si_2O_{13}$), plus an Al–Si spinel. At temperatures above 1000°C and at relatively long heating durations the silica polymorph cristobalite begins to form.

Firing of kaolinite produces a pure white product, which consists predominantly of mullite. Mullite will also be formed by firing other clay minerals, but with the introduction of additional compositional components, firing temperatures will be reduced and other mineral phases may be produced. To some extent, these can be predicted hypothetically from equilibrium phase diagrams. Figure 8.10 shows solid mineral phase assemblages which are stable at high temperature within the systems SiO_2–Al_2O_3–MgO, SiO_2–Al_2O_3–CaO and SiO_2–Al_2O_3–K_2O, together with plots of the unstable clay minerals clinochlore, illite, kaolinite and smectite. In

MINERALOGICAL CHANGES DURING FIRING

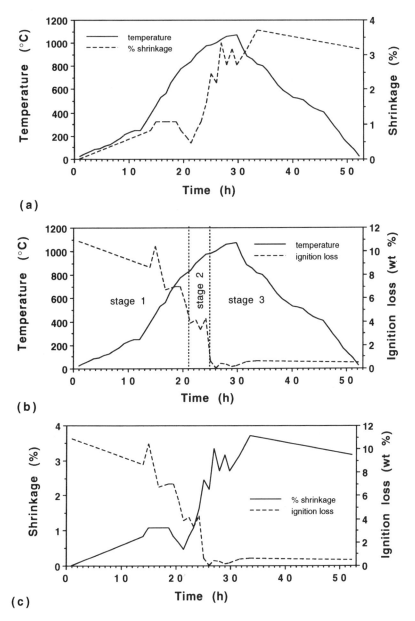

Figure 8.8 (a) an example of a firing curve for a batch of fireclay–shale bricks, showing the shrinkage measured on samples removed from the kiln at intervals along its length during firing. (b) shows the ignition loss of the same brick samples, identifying an initial stage of decreasing ignition loss, a second stage of ignition loss remaining constant at about 4 wt%, and a third stage of low residual ignition loss. (c) allows comparison of the shrinkage and ignition loss curves. Drawn from data collected by J.C. Thomas.

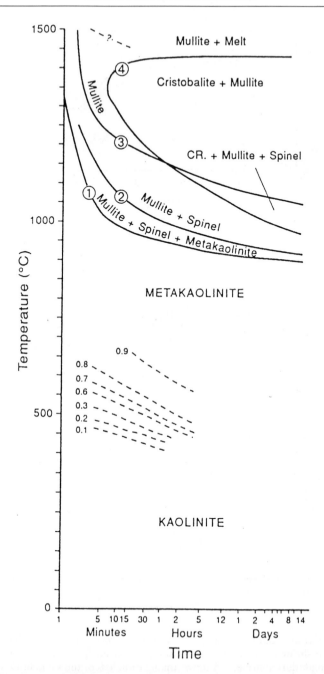

Figure 8.9 Time–temperature–transformation (TTT) diagram for the firing of kaolinite (from Dunham, 1992). Solid lines indicate boundaries between mineral stability fields, and the dashed lines indicate the weight fraction of OH lost during the change from kaolinite to metakaolinite.

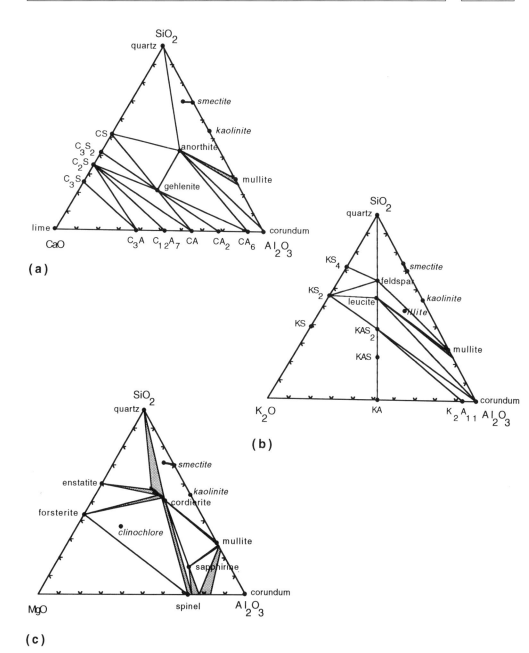

Figure 8.10 Ternary diagrams, with compositions expressed in weight %, for the systems (a) SiO_2–Al_2O_3–CaO, (b) SiO_2–Al_2O_3–K_2O and (c) SiO_2–Al_2O_3–MgO, showing the positions of the clay minerals (smectite with a range in composition, illite and kaolinite) in italics and the high-temperature minerals composed of these oxides. In Figure 8.10(b) 'feldspar' denotes orthoclase.

theory, if fired at 1000–1100°C, clinochlore will yield an assemblage consisting of forsterite (the Mg olivine), spinel and cordierite, whereas illite will yield mullite, orthoclase feldspar and leucite (depending on the illite composition). Smectite will yield an assemblage dominated by mullite and a silica mineral (cristobalite, tridymite or quartz, depending on temperature), anorthite and cordierite, depending in detail on the smectite composition.

Of particular interest in designing firing conditions is the behaviour of iron. If bricks are fired under oxidizing conditions, any free iron oxide that is present will form haematite, as a finely dispersed red oxide which gives a red colour to the brick. If reducing conditions are used, other iron-bearing phases form in which the iron is present as Fe^{2+}, such as ilmenite or ferrous silicate minerals (e.g. iron cordierite), and this results in a blue brick. Figure 8.11 illustrates in principle the way in which the amount of oxygen present during firing, measured in terms of the fugacity of oxygen, influences the relative stability of haematite and magnetite. The shape of the haematite–magnetite reaction curve (and other curves relating to other oxygen-buffering mineral reactions) is such that an increase in temperature (the manufacturing procedure of 'flashing') will stabilize reduced phases – higher temperature bricks will be blue, low temperature red. Alternatively, starvation of the kiln of oxygen at constant temperature, by reducing the amount of air available for combustion, has a similar effect by reducing

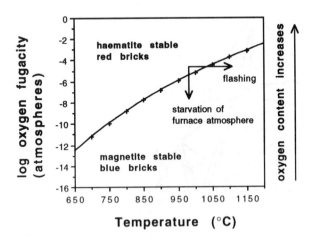

Figure 8.11 Stability of haematite and magnetite as a function of temperature and oxygen fugacity (calcuated using data from Eugster and Wones, 1962). The effects of extremes of flashing and starvation of the furnace of air are shown by arrows crossing the boundary between the haematite and magnetite stability fields; a combination of both procedures would give an intermediate trajectory.

the oxygen fugacity again allowing the reaction curve to be crossed. In practice, a combination of the two procedures is often used, giving an intermediate trajectory of decreasing oxygen fugacity and increasing temperature. This phenomenon can also be used to explain the colour difference between core and margins of bricks made from clays with high organic matter (or pyrite) contents, where reduction may take place in the cores, giving a blue or grey core, and a red exterior. It should, however, be noted that although haematite is a common constituent of bricks magnetite is not, but reduced iron is contained in minerals such as ilmenite and silicates. Pale coloured bricks are produced either from bulk clays with a low iron content (such as fireclays or ball clay), or from clays in which there is sufficient calcium (particularly derived from calcite) to result in the production of calcium–iron silicate minerals. In particular, if Ca predominates over Fe, buff or cream bricks are produced under normal conditions. If fired under reducing conditions, high calcium clays yield yellow bricks. In addition, a proportion of iron (up to 10% Fe_2O_3) may be accommodated by solid solution within mullite, again preventing the formation of the red-brown haematite pigment.

8.3 MINERALOGY OF BRICKS

In practice, bricks are complex mineralogically, and because of their very fine grain-size it can be difficult to characterize adequately the phases that are present. The situation is complicated by the fact that a brick will consist of relatively large fragments of rock and quartz, which have generally not participated in the firing reactions, surrounded by a matrix of products of the firing of the clay mineral component. Plates B(ii), B(iii) and B(iv) and Figures 8.12 and 8.13 show petrographically the variation in texture typical of bricks made from ground shale or till.

Note that the use of reflected light optics to study polished thin sections of bricks is valuable (although unorthodox), as it reveals textural information without misleading effects caused by the grains being smaller than the thickness of the petrographic thin section. Scanning electron microscopy has the advantage of providing compositional information, if X-ray analysis facilities are available, but very often the grain-size of the brick is too fine to allow individual grains to be resolved. As well as emphasizing the heterogeneity which is typical of many bricks, Figures 8.12 and 8.13 show the vesicular texture and the large proportion of pore space, produced by degassing during firing.

Detailed studies of brick mineralogy have identified a number of dominant minerals (Dunham, 1992): quartz, cristobalite, mullite, potassium feldspar, plagioclase, wollastonite, pyroxene, melilite, anhydrite, haematite and glass. The formation of these minerals is summarized

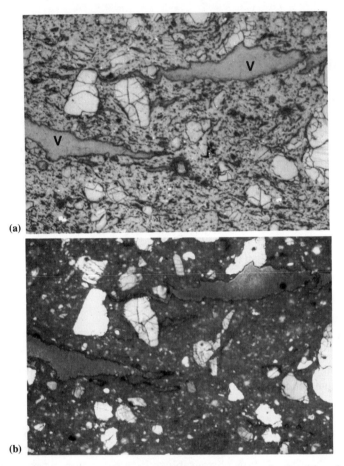

Figure 8.12 Photomicrographs of polished thin sections of a boulder clay brick, showing quartz grains set in fine-grained matrix, with large voids (v) and fine scattered small voids (dark specks throughout (a)). Photograph (a) taken using reflected plane polarized light; (b) viewed under crossed polars (internal reflections enable birefringent grains to be clearly visible). The field of view measures 2.85 × 1.85 mm.

graphically in Figure 8.14, which shows the loss of certain components of the bulk clay and the new minerals which are formed from them.

8.4 ASSESSMENT OF BRICK CLAY RAW MATERIALS

In the light of the complexity of brick mineralogy, it is perhaps not surprising that an empirical approach is taken in the selection and blending of brick clays. However, brick clay reserves are under pressure, and a more sophisticated approach will become increasingly appropriate (Prentice,

ASSESSMENT OF BRICK CLAY RAW MATERIALS

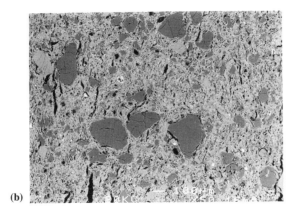

Figure 8.13 Scanning electron micrographs (back-scattered electron images) of polished thin sections of bricks. The field of view measures 2.85 × 1.85 mm. (a) fired Etruria Marl; note rock fragments (approximately 1 mm in diameter) and quartz sand grains (approximately 0.1 mm in diameter) set in a matrix of fired clay; (b) fired Boulder Clay; note quartz fragments (approximately 0.2 mm in diameter), set in a matrix of fired clay. Black areas are shrinkage cracks.

1988). The bulk composition of the clay, as determined by X-ray fluorescence, can be taken as a starting point, and one clay compared with another accordingly. However, a number of steps need to be taken in interpreting the data shown in Table 8.1. First, each analysis includes a value reported for the 'loss on ignition' (LOI), which represents the volatile matter that is lost on heating at a high temperature, usually 1000°C or 1100°C. This laboratory test directly parallels ignition which takes place during firing. The gases evolved include water and carbon dioxide, but also sulphur dioxide if sulphides are present in the clay. During brick manufac-

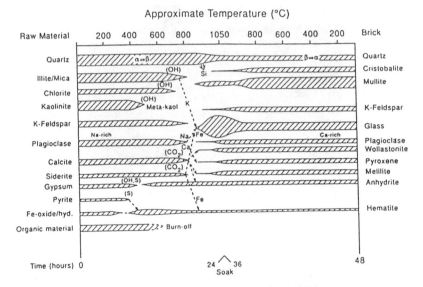

Figure 8.14 Summary of mineralogical changes which take place during firing involving a maximum temperature of 1050°C (adapted from Dunham, 1992).

ture these gases escape from their mineral hosts, giving the porous texture shown in Figures 8.12 and 8.13.

The first step in comparing chemical analyses is to remove the volatiles represented by the LOI, so that compositions which remain in the brick can be compared. This is done by recalculating the non-volatile oxides to 100%, as has been done in Table 8.2. Some of the apparent differences between samples have now diminished, as the diluting effect of the volatile component has been removed. It is then convenient to compare compositions in ternary diagrams which contain phase compatibility data, so that the mineralogy of the finished brick can be predicted, bearing in mind that the use of a number of three-component diagrams is merely an approach to describing the behaviour of the complex multicomponent systems of interest. Table 8.2 includes the parameters needed for plotting the compositions in the ternary diagrams SiO_2–Al_2O_3–CaO, SiO_2–Al_2O_3–FeO, SiO_2–Al_2O_3–K_2O and SiO_2–Al_2O_3–MgO, which are presented in Figure 8.15.

Almost all of the clay compositions plot within compatibility triangles bounded by the silica mineral–mullite join, consistent with the importance of this assemblage as a component of bricks. The third phase depends on the diagram used, but includes cordierite (Fe- and Mg-bearing systems), anorthite (plagioclase; Ca-bearing system) or potassium feldspar (K-bearing system). Some bulk compositions lie outside this triangle, and these will be clays in which minerals other than mullite might be produced

Table 8.2 Compositions of brick clay raw materials (from Table 8.1), recalculated as normalized, volatile-free compositions (wt %) and as ternary proportions

Component	1 Shale	2 Fireclay	3 Etruria Marl	4 Etruria Marl	5 Mercia Mudstone	6 Mercia Mudstone	7 Oxford Clay	8 Boulder Clay
SiO_2	53.69	56.30	58.12	51.39	48.70	65.30	45.82	63.41
TiO_2	0.20	1.16	1.35	1.27	0.69	0.74	0.84	0.70
Al_2O_3	20.50	25.50	22.40	23.10	13.30	14.00	15.21	13.81
Fe_2O_3	6.95	3.19	6.79	9.02	5.05	4.55	2.81	4.57
FeO	0.86	–	–	–	–	–	–	–
CaO	0.30	0.30	0.40	2.00	5.59	1.69	10.10	2.40
MgO	2.41	0.74	1.28	0.86	8.37	2.73	2.21	3.27
Na_2O	0.62	0.11	0.14	0.10	0.09	0.53	0.86	–
K_2O	2.73	2.13	1.65	1.79	5.07	5.03	2.62	2.34
SO_3	0.37	–	–	–	–	–	–	–
LOI	11.14	9.76	7.40	9.09	11.80	4.48	15.00	8.70
Totals	99.77	99.19	99.53	98.62	98.66	99.05	95.47	99.21
Normalized volatile-free compositions:								
SiO_2	60.76	62.95	63.08	57.40	56.07	69.05	56.94	70.06
TiO_2	0.23	1.30	1.47	1.42	0.79	0.78	1.04	0.78
Al_2O_3	23.20	28.51	24.31	25.80	15.31	14.80	18.90	15.26
Fe_2O_3	8.95	3.57	7.37	10.07	5.81	4.81	3.49	5.05
CaO	0.34	0.34	0.43	2.23	6.44	1.79	12.55	2.65
MgO	2.73	0.83	1.39	0.96	9.64	2.89	2.75	3.61
Na_2O	0.70	0.12	0.15	0.11	0.10	0.56	1.07	0.00
K_2O	3.09	2.38	1.79	2.00	5.84	5.32	3.26	2.59
Totals	100.00	100.00	100.00	100.00	100.00	100.00	100.00	100.00
ΣFe as FeO	8.06	3.21	6.63	9.07	5.23	4.33	3.14	4.54
Ternary proportions:								
SiO_2	66.03	66.49	67.09	62.21	73.18	78.30	72.09	77.96
Al_2O_3	25.21	30.12	25.86	27.96	19.99	16.79	23.93	16.98
FeO	8.76	3.39	7.05	9.83	6.83	4.91	3.98	5.06
SiO_2	72.08	68.57	71.82	67.19	72.05	80.63	64.42	79.64
Al_2O_3	27.52	31.06	27.68	30.20	19.68	17.29	21.38	17.35
CaO	0.40	0.37	0.49	2.61	8.27	2.09	14.20	3.01
SiO_2	70.09	68.21	71.05	68.20	69.21	79.61	72.45	78.78
Al_2O_3	26.76	30.89	27.38	30.66	18.90	17.07	24.05	17.16
MgO	3.15	0.90	1.56	1.15	11.89	3.33	3.49	4.06
SiO_2	69.80	67.08	70.73	67.37	72.61	77.43	71.99	79.70
Al_2O_3	26.65	30.38	27.26	30.28	19.83	16.60	23.90	17.36
K_2O	3.55	2.54	2.01	2.35	7.56	5.96	4.12	2.95

1, Weeton Shales, Namurian, Lancashire/Yorkshire (includes 0.13% FeS_2; Ridgeway, 1982); 2, Etruria Marl, Staffordshire (Ridgeway, 1982); 3, Calcareous Etruria Marl, Staffordshire (Ridgeway, 1982); 4, Fireclay, Upper Carboniferous, Leicestershire; 5, Mercia Mudstone (Keuper Marl), Triassic, Leicestershire (Ridgeway, 1982); 6, Mercia Mudstone (Keuper Marl), Triassic, Leicestershire (Ridgeway, 1982); 7, Lower Oxford Clay, Jurassic, Fletton, Northamptonshire (includes 1.83% FeS_2; Ridgeway, 1982); 8, Quaternary Boulder Clay, Cheshire.

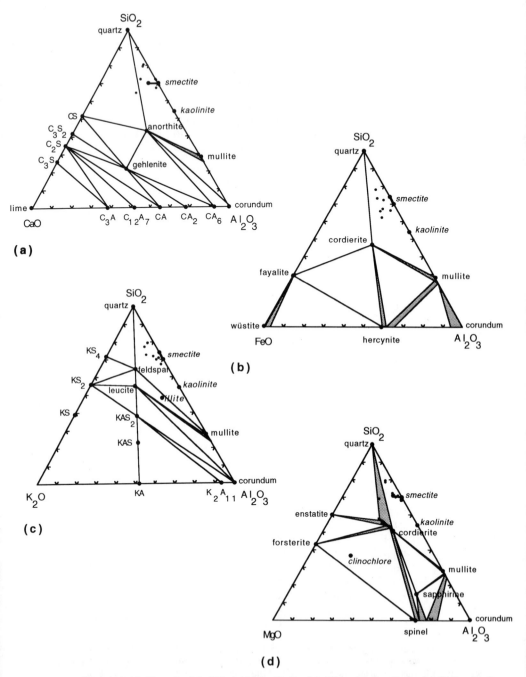

Figure 8.15 Plots in (a) SiO_2–Al_2O_3–CaO, (b) SiO_2–Al_2O_3–FeO, (c) SiO_2–Al_2O_3–K_2O and (d) SiO_2–Al_2O_3–MgO of typical brick clay bulk compositions, expressed as weight %.

if homogeneous firing is achieved. For example, the Oxford Clay (sample 7 in Table 8.2) has a high CaO content, which results in its plotting close to the silica mineral–anorthite join in the triangle CS–silica mineral–anorthite (Figure 8.15a); this clay might be predicted to produce calcium silicates rather than mullite on firing. Indeed, this is the case, and pale bricks are produced as a consequence of the formation of the calcium silicate minerals, which also take up iron preventing the formation of haematite. Similarly, the Mercia Mudstone (Keuper Marl) is potentially (but by no means always) rich in MgO (sample 5 in Table 8.2), plotting in the silica mineral–cordierite solid solution field in Figure 8.15d. This indicates that magnesium silicates will form, which again accommodate iron and favour a pale brick. Although many bulk compositions of the clays plot within the triangle with mullite and a silica mineral at two of its corners, individual mineral components (such as clinochlore and other non-clay minerals) may plot in other fields, and calcite plots at the CaO apex. The lack of homogenization during the firing process allows a number of mineral assemblages to be developed, as a consequence of local variation in composition of reactive components as well as localized reaction between the clay, calcite and rock fragment components of the brick.

The diagrams shown in Figure 8.15 show no information relating to the temperatures involved in the production of the glass phase which binds the brick together. For the systems $SiO_2–Al_2O_3–CaO$, $SiO_2–Al_2O_3–FeO$ and $SiO_2–Al_2O_3–MgO$, the minimum in the silica-rich part of the diagram is well above widely-used brick firing temperatures (i.e. above 1100°C), and so these diagrams cannot be used to predict melting behaviour during production. However, in the system $SiO_2–Al_2O_3–K_2O$, the liquidus surface rapidly descends with increasing K_2O content to temperatures as low as 985°C for the composition 80% SiO_2, 11% Al_2O_3 and 9% K_2O. In partial melting at this temperature, this would be the composition of the first-formed melt, emphasizing the importance of the clay potassium content to the brick manufacturer. Figure 8.16 shows a sketch of an isothermal section at 1100°C for the part of the system $SiO_2–Al_2O_3–K_2O$ which relates to bulk compositions within the mullite-silica mineral-orthoclase field. It shows fields for liquid, liquid + tridymite, liquid + tridymite + mullite and liquid + mullite. Hypothetical bulk compositions for brick clays are denoted by points a and b, which have been drawn to lie on a straight line connecting the liquid composition, l, and the mullite–tridymite join. Composition a is more potassium rich than composition b, and, using the lever rule, it will contain more liquid (approximately 60%) than b (approximately 50%). If a similar isothermal section is produced for a lower temperature, point l will move away from the mullite–tridymite join, and the proportions of melt in each case will decrease, so that in the example drawn (at 1000°C) in Figure 8.16, composition a now consists of

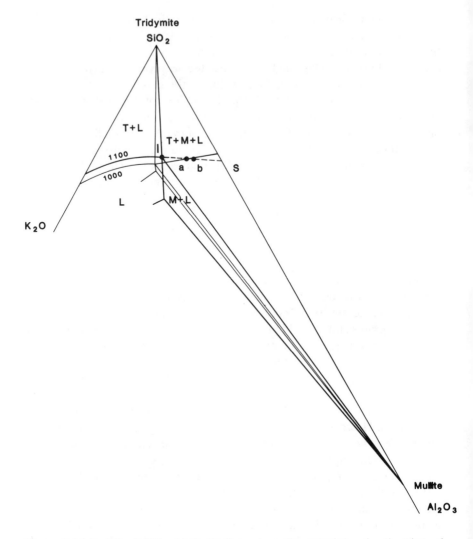

Figure 8.16 Detail of SiO_2–Al_2O_3–K_2O to show the principle of estimation of proportions of melt present at 1000°C and 1100°C, using isothermal sections. Points a and b refer to bulk compositions discussed in the text (expressed in weight %), which at 1100°C consist of a mixture of liquid 1 and solid assemblage s (tridymite + mullite). Note that tridymite is the stable silica polymorph at this temperature. T = tridymite, M = mullite and L = liquid.

approximately 50% liquid. If we suppose that brick manufacture involves a delicate balancing act between producing just enough melt to bind the brick together but not so much that the brick loses shape, we can suggest that for an 'ideal' brick there may be an ideal proportion of melt formed,

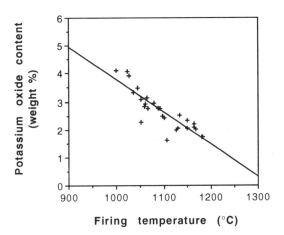

Figure 8.17 Firing temperatures for a suite of Carboniferous shales as a function of their potassium oxide contents (from information supplied by C.D. Curtis and D. Ashby).

and that that proportion is more or less constant irrespective of the bulk clay composition. It follows from the isothermal construction shown in Figure 8.16 that to maintain constant proportions of liquid the temperature of firing must be varied according to the composition of the clay. In this hypothetical case, to achieve 50% melting, composition b must be held at 1100°C and composition a at 1000°C. This is necessarily an oversimplified view, but it illustrates the principle behind the practice that different clays (or blends of clay) are fired at different temperatures. Indeed, there is a close correlation between firing temperature and bulk clay potassium contents (Figure 8.17). Despite the value of phase diagrams in explaining the scientific background to industrial practices, brick firing temperatures are generally determined by experience rather than by the predictive use of phase diagrams.

In addition to occurring within the smectite clays, lime (CaO) can occur as calcium carbonate (calcite or limestone) or within gypsum. If present, it is important to determine whether or not the calcium carbonate is finely disseminated, in the form of fragments of limestone, or as concretions. Finely disseminated calcium carbonate is not detrimental to brick firing, and indeed may be beneficial (by reacting with sulphur dioxide given off as sulphides decompose, so reducing flue gas emissions). In contrast, visible fragments of limestone or concretions must be avoided because of the way in which they cause the brick surface to spall off (as described earlier). This defect is commonly observed in areas where tills containing limestone fragments have been used as a brick clay.

8.5 ENVIRONMENTAL ASPECTS OF BRICK PRODUCTION

So far, we have concentrated on the chemical processes involved in the transformation of a brick clay into a fired product. Brick manufacture has associated with it a number of environmental implications, some beneficial, others potentially detrimental.

Traditionally, bricks have been and are produced in factories which are sited adjacent to the source of the clay. The life of the production site may extend beyond the life of an associated quarry, or beyond individual phases of quarrying, yielding large excavations in the vicinity of the plant. These are ideal for consideration as waste disposal sites, as the 'bedrock' is clay, with a low permeability, and with little or no upgrading may meet the requirements of the waste disposal authorities (Chapter 11). The benefits to the brick producer are enormous. First, it may be possible to sell the hole for more than it cost to excavate, in effect making the extraction of brick clay a by-product of preparing a site for a disposal operation. Secondly, a domestic waste disposal site represents a valuable source of fuel in the form of the landfill gas generated as the waste decomposes (Chapter 11), and if recovered for the brick factory this can greatly ease the costs of firing. Thus brick manufacturing and waste disposal can in many circumstances be regarded as a 'symbiotic' process.

A second area where the manufacture of bricks has an impact on environment management lies in the emissions produced during firing, and ensuring that these are minimized. The loss on ignition term of the chemical analysis includes not only water and carbon dioxide but also harmful species such as sulphur dioxide, chlorine and fluorine, which escape as acid gases during firing unless mineralogical reactions retain them in the brick. The presence of finely disseminated lime, for example, will react with sulphur dioxide and fluorine to produce anhydrite and fluorite, preventing the escape of these gases. Other components of the brick also interact with the gases, and additives (such as barium carbonate) can be purchased for blending with the raw material prior to firing. In an assessment of the potential to generate acid gases, it is important to assess the composition not only of the raw materials but also the fired products, to ensure that their fate during firing is known.

Refractories 9

Earlier in this book, we considered the use of industrial minerals as raw materials for the manufacture of a number of products whose processing requires high temperatures. We have paid little attention to the very special problems of containment that the manufacturing processes involve, which include contact with often corrosive solids, melts and gases at high temperatures. These problems are solved by the use of refractory materials as linings to furnaces, kilns, flues, reaction vessels and ladles. Refractories are carefully selected to be compatible with the conditions of service, and must meet the following ideal requirements:

- they must be physically capable of performing under the conditions of the process, without failure of any kind. They must not melt, and they must not fracture or spall in response to thermal stresses.
- they must be chemically capable of performing under the required conditions – they must not react with the materials which are being processed, or dissolve within them.

These conditions may well not be met in practice, requiring periodic replacement of refractories and postmortem analysis of those which have failed or show signs of wear. For many reasons, refractory linings are usually in the form of bricks rather than a single coating; their compositions can be matched to suit the conditions in particular zones within a kiln, they can be replaced or installed piecemeal, and they can be manufactured to suit particular geometries. There is a wide variety of materials available for refractory purposes, and only a brief introduction is presented here. For more detailed information, the textbook by Chesters (1983) and the Industrial Minerals Consumer Survey (Dickson, 1986) should be consulted.

9.1 CONDITIONS OF SERVICE

High temperatures, up to 1500°C or more, are required for glass, cement and steel manufacture. Under these conditions (and in the presence of

Table 9.1 Mineral refractories and their melting points

Mineral refractory	Melting point, °C
Fireclay brick	1600–1750
Silica brick	1700
Bauxite brick	1730–1850
High alumina clay brick	1802–1880
Mullite	1810
Sillimanite	1816
Forsterite	1890
Chromite	1770
Chrome brick	1950–2200
Alumina	2050
Spinel	2135
Magnesite brick	2800
Zirconia brick	2200–2700
Silicon carbide	2700
Boron nitride	2720

corrosive liquids and gases) most metals fail, although refractory metals can be used for small scale operations. In contrast, mineral-based refractories provide a number of possible materials, depending in detail on the conditions. A list of some mineral refractories and their melting points is given in Table 9.1.

Note that some of the materials given in Table 9.1 are pure minerals in their own right (silica, mullite, sillimanite, forsterite, chromite, corundum, spinel and periclase), whereas others are composite materials made up of many minerals (fireclay brick) or materials which are not known as minerals (zirconia, silicon carbide, boron nitride). The stability in service of refractories can to some extent be predicted using phase diagrams. The melting points given in Table 9.1 correspond to the liquidus temperatures for the refractory minerals in the appropriate phase diagrams, and reactions with melts can be predicted by inspection of phase diagrams in which both the melt of interest and the refractory mineral can be plotted.

9.2 SILICA REFRACTORIES

Silica refractories (Highley, 1977) are simple chemically, composed of SiO_2, but they are more complex mineralogically. There are several polymorphs of silica, including α-quartz, β-quartz, tridymite and cristobalite as the relevant low pressure polymorphs. Their stability depends on temperature (Table 9.2). Other polymorphs exist but are metastable.

If pure silica is heated to its melting point it will change from one polymorph to another, as appropriate for the temperature, provided there

Table 9.2 Silica polymorphs: their thermal stability and densities

Polymorph	Thermal stability (°C)	Density
α-quartz	up to 573	2.65
β-quartz	573–870	2.65
β-tridymite	870–1470	2.26
β-cristobalite	1470–1713	2.33

is enough time. Because these silica minerals are composed of three dimensional polymeric networks of Si and O, rearrangement of those networks is slow kinetically, and metastability is an important phenomenon in human terms. Of course, over geological periods of time inversion (the transformation on cooling from one polymorph to another stable at lower temperatures) to α-quartz takes place.

Chemically, silica bricks are unreactive in dry circumstances. They become increasingly reactive with melts as the basicity of the melt increases – a consequence of the reaction to form ferromagnesian silicates, with which geologists are familiar:

$$\text{forsterite} + \text{silica} = \text{enstatite} \tag{9.1}$$

This can be illustrated in Figure 9.1, which shows phase relationships for the two-component system MgO–SiO_2. A hypothetical basic slag composition can be simplified to allow it to be plotted within this system, and if it lies in the field of stability of the two-phase assemblage periclase + melt (e.g. composition X in Figure 9.1) it will react with silica to yield first enstatite and then forsterite, forming as skins on the solid part of the brick (Figure 9.2), and as reaction rims around individual grains as the slag penetrates the pore space.

This reaction means that silica bricks are unsuitable for applications involving basic slags, although they may be used in circumstances where they do not come into contact with such slags but come into contact with hot gases (e.g. coke furnace linings, blast furnace air flue linings etc.).

Physically, silica bricks are very stable. Variation in the linear thermal expansion coefficient varies from one polymorph to another (see Figure 8.6), with sharp changes as temperature increases for quartz and cristobalite. Tridymite differs from the other two polymorphs by having an even variation in thermal expansivity. Thus when tridymite is heated or cooled, there are no sudden changes in coefficient of thermal expansion, meaning that mechanical stresses during heating and cooling are minimized, allowing tridymite bricks to maintain their integrity.

To obtain bricks with a high proportion of tridymite, quartz is fired at 1450°C for long periods, until the density becomes 2.3–2.35. They need to

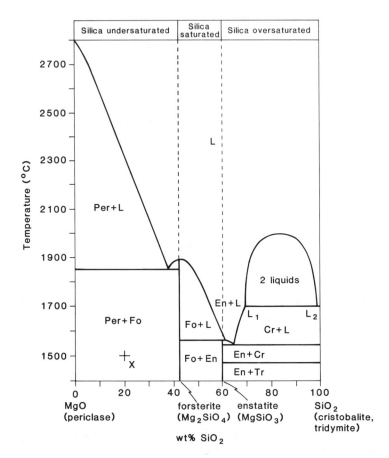

Figure 9.1 Phase relationships for the binary system MgO–SiO_2 at one atmosphere pressure. X marks the composition of the slag whose behaviour is discussed in the text, expressed in weight %.

be low in porosity, to prevent liquids or gases from penetrating the interior of the brick. The end products are rarely completely tridymite, as the conversion of quartz is not 100% efficient. Impurities such as Al reduce the refractory quality of the bricks.

Finished silica bricks do not shrink at temperatures up to their melting point (1710°C), can be used under load up to 1650°C and are resistant to thermal shock. Their life can be prolonged by avoiding attack by basic melts, and by maintaining them at temperatures above about 700°C throughout their working life.

The properties that make silica suitable for refractory bricks also favour its use for foundry work, which is dominated by the use of silica sands as moulding sands. Moulding sands differ from glass sands (see Chapter 6) by

MAGNESIA REFRACTORIES

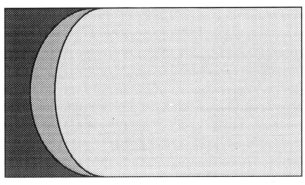

Altered silica brick   Unaltered silica brick

Figure 9.2 Schematic diagram to show reaction geometry of silica brick reacting with basic slag.

being less pure mineralogically and chemically, within limits. Mineralogical 'impurities' such as clays act as natural binders and are acceptable or desirable. However, silica sand of high purity is also used, with resins, added bentonite or other chemicals as binding materials. Of particular importance is grain-size distribution, to facilitate mould manufacture and to provide a desired surface finish for the east article.

Geological sources of quartz suitable for the manufacture of silica bricks are widespread. In addition to unconsolidated sands, pure sandstones (gannisters) can be quarried and fired after crushing. Bricks made from crushed rock frequently show a wide range of grain sizes when examined petrographically.

9.3 MAGNESIA REFRACTORIES

Magnesia refractories are simple chemically, being composed essentially of MgO, known mineralogically as periclase. Periclase has a specific gravity of 3.56–3.68 and melts at about 2800°C. Being basic, it can be used for applications where silica bricks are unsuitable, and it can also be used for higher temperature work. It is used, for example, in basic oxygen furnaces in steelmaking, and as a lining in the burning zone of cement kilns.

Again, the stability of periclase-based bricks in service can be predicted, at least in part, using the phase diagram for the system $MgO–SiO_2$ given in Figure 9.1. In this example, a periclase brick will be chemically stable in the presence of a basic slag of composition X. However, although this means that no new crystalline phases are expected to form by reaction between the slag and the brick, the brick may become corroded through partial dissolution into the slag. The stability of magnesia refractories is discussed in more detail later, when slag chemistry is examined more closely.

There are two dominant raw materials which are used to prepare magnesia (periclase) refractories – dolomite and magnesite. Magnesite is the naturally occuring magnesium carbonate, with the formula $MgCO_3$. Dolomite, however, is a frequently misused and often poorly defined term (O'Driscoll, 1988). It refers both to the mineral $CaMg(CO_3)_2$, and to the rock made up of that mineral, which often contains other mineral impurities. Care should be taken to define clearly the use of the term in practice.

Both dolomite (mineral or pure dolomite rock) and magnesite are converted to magnesia by calcining to remove CO_2, which involves heating to about 1500°C. However, the yield from dolomite mineral includes CaO, and in order to produce pure magnesia a subsequent process is used, involving seawater.

Seawater naturally contains the equivalent of 0.2% MgO, and $Mg(OH)_2$ is less soluble than the equivalent Ca compound $Ca(OH)_2$. $Mg(OH)_2$ can therefore be precipitated preferentially under alkaline conditions.

Seawater is initially treated with acid to remove CO_2 which is naturally present. It is then treated with slaked 'dolime' (slaked calcined dolomite rock, a mixture of $Ca(OH)_2$ and $Mg(OH)_2$) which raises the pH so that magnesium hydroxide precipitates onto previously prepared seed crystals. The precipitated magnesium hydroxide is then filtered off and calcined to give dead-burned magnesia, in the form of pellets. In Britain, this process is carried out in north-east England, where the outcrop of the Permian Magnesian Limestone is intersected by the sea.

Once produced, dead-burned magnesia pellets are used to produce refractory bricks – but care has to be taken to avoid the pellets or bricks becoming damp, which would slake the magnesia. Consequently, magnesia bricks are impregnated with pitch or oil, or increasingly nowadays are manufactured with a more sophisticated graphite or resin binder.

9.4 ALUMINOSILICATE REFRACTORIES

In many applications, the use of magnesia or silica refractories may not be appropriate, because of reaction or physical stability problems as well as on the grounds of cost. Aluminosilicate refractories represent a third major compositional group, which includes refractories produced from clays as well as from pure aluminium silicate minerals. Their stabilities can be deduced from the phase diagram for the binary system Al_2O_3–SiO_2 (Figure 9.3), which forms part of the ternary systems considered in Chapter 8. In this system, the types of refractory that can be produced are related to the

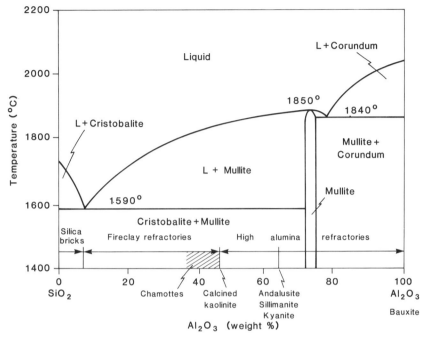

Figure 9.3 Phase relationships for the binary system Al_2O_3–SiO_2 (weight %) at one atmosphere pressure, showing the compositional range of materials used to make aluminosilicate refractories.

raw materials from which they are made, with their alumina content increasing (and their silica content decreasing) in the following order:

	$SiO_2:Al_2O_3$ molecular ratio	$Al_2O_3:SiO_2$ molecular ratio
fireclay	>2	<0.5
chamotte or kaolin	2	0.5
sillimanite, andalusite	1	1
bauxite	0	≫1

As considered in Chapter 8, the mineral assemblages produced on calcining kaolinite rich clays will include mullite and a silica mineral, and for alumina compositions in excess of 72% Al_2O_3, corundum will be stable with mullite.

9.4.1 Clay-based refractories

The use of clays in the production of refractories has a long history, taking advantage of the widespread occurrence and ease of availability of clay raw materials, which when fired yield bricks suitable for structural work or containment of materials where contamination is not expected to be a problem. A number of different clay raw materials are involved.

(a) Fireclays

Traditionally very extensively used, fireclays commonly occur in association with coal bearing sequences, and can be produced as a by-product of a coal mining operation. Of the refractories listed in Table 9.1, fireclay bricks have the lowest melting points, limiting their use to the least demanding circumstances. They are clays which are dominated by kaolinite and illite, with alumina contents varying between 25% and 40%. Impurities include quartz and titanium oxide minerals, and their silica contents are in excess of those expected for mixtures of kaolinite and illite (Table 9.3). Fireclay refractories are declining in importance; they are mainly used for structural work, in support of or behind refractories with higher performance specifications.

(b) Flint clays

These are non-plastic, kaolinite-rich, clays which differ from other sedimentary kaolins (Chapter 3) in terms of their hardness and relatively high density. They break with a conchoidal fracture, hence their name, and have a low firing shrinkage (around 12%). Again, kaolinite predominates, with minor quantities of illite and quartz. The flint clays which are most sought after are those which also contain aluminium oxides such as diaspore or boehmite, which naturally increase their alumina content from the already high 35–40% Al_2O_3. Flint clays are an important product in the United States, and may be sold calcined or as mined.

(c) Chamotte

Because of the natural scarcity of flint clays, some aspects of their characteristic properties can be reproduced by calcining a more widely available plastic clay (such as a ball clay) to form a non-plastic, hard, dense, refractory aggregate. This is common practice outside North America, and the product obtained is known as chamotte, refractory grog or calcined kaolin. The major European producers of chamotte are France (producing material in the range 37–48% Al_2O_3 and Germany (with products up to about 40% Al_2O_3). In Britain, calcined fireclays are produced from Carboniferous sequences, and calcined kaolinite ('molochite') is produced from the south-west England kaolin deposits.

Table 9.3 shows compositions of a number of kaolin-rich clays which are suitable for refractory purposes, together with the compositions of pure

Table 9.3 Compositions of refractory clays and related materials (wt %)

Component	1 kaolinite	2 illite	3 china clay	4 fireclay	5 ball clay 1	6 ball clay 2	7 bauxitic clay	8 flint clay	9 chamotte
SiO_2	46.55	45.08	46.20	54.20	48.00	67.00	39.30	44.42	–
TiO_2	0.00	0.00	0.09	1.25	0.90	1.40	4.89	2.12	–
Al_2O_3	39.50	38.63	39.20	29.30	34.00	22.00	35.70	38.63	–
Fe_2O_3	0.00	0.00	0.23	2.10	1.00	0.90	4.19	0.55	–
CaO	0.00	0.00	0.06	0.27	0.20	0.10	0.39	0.04	–
MgO	0.00	0.00	0.07	1.06	0.30	0.30	0.29	0.10	–
Na_2O	0.00	0.00	0.09	0.09	0.20	0.30	0.14	0.12	–
K_2O	0.00	11.78	0.21	2.94	1.60	2.20	0.10	0.30	–
LOI	13.95	4.50	13.80	8.53	13.80	5.80	15.00	13.90	–
Totals	100.00	99.99	99.85	99.74	100.00	100.00	100.00	100.18	–
Expressed on an anhydrous basis:									
SiO_2	54.10	47.21	53.69	59.42	55.68	71.13	46.24	51.48	57.40
TiO_2	0.00	0.00	0.10	1.37	1.04	1.49	5.75	2.46	2.00
Al_2O_3	45.90	40.45	45.55	32.12	39.44	23.35	42.00	44.77	37.00
Fe_2O_3	0.00	0.00	0.27	2.30	1.16	0.96	4.93	0.64	1.80
CaO	0.00	0.00	0.07	0.30	0.23	0.11	0.46	0.05	0.35
MgO	0.00	0.00	0.08	1.16	0.35	0.32	0.34	0.12	0.35
Na_2O	0.00	0.00	0.10	0.10	0.23	0.32	0.16	0.14	0.10
K_2O	0.00	12.34	0.24	3.22	1.86	2.34	0.12	0.35	0.85
Totals	100.00	100.00	100.00	100.00	100.00	100.00	100.00	100.00	99.85
Ternary compositions in SiO_2–Al_2O_3–K_2O:									
SiO_2	54.10	47.21	53.97	62.70	57.42	73.46	52.33	53.29	60.26
Al_2O_3	45.90	40.45	45.79	33.90	40.67	24.12	47.54	46.35	38.85
K_2O	0.00	12.34	0.25	3.40	1.91	2.41	0.13	0.36	0.89

1, kaolinite; calculated composition for $Al_2Si_2O_5(OH)_4$; 2, illite; calculated composition for $KAl_3Si_3O_{10}(OH)_2$; 3, china clay (kaolinite, St. Austell; Deer, Howie and Zussman, 1992); 4, fireclay, south Staffordshire coalfield, England (from Table 11, Highley, 1982); 5, ball clay (Group 1 clays, Devon; Watts Blake Bearne technical literature); 6, ball clay (Group 4 clays, Devon; Watts Blake Bearne technical literature); 7, bauxitic clay, Ayrshire, Scotland (recalculated from Table 4 in Highley, 1982); 8, flint clay, Missouri (from Bristow, 1989); 9, chamotte, commercial product Sarca 35V (Dickson, 1986).

kaolinite and muscovite. Note that correction must be made for the water content of the clay, and many analyses in the literature (academic or technical) are reported for calcined material, which has already been fired prior to analysis and so can be considered on an anhydrous basis. Of the clay minerals, kaolinite yields the highest potential alumina content, up to a maximum of approximately 46% Al_2O_3 for the calcined pure clay. These values are only approached for the purest natural kaolins, such as the flint clays. However, a manufacturer of aluminosilicate refractories typically produces a range of products with differing alumina contents, permitting the use of a number of different clay raw materials. Of particular importance is the use of bauxite or bauxite-rich clays as a means of producing high alumina refractories. Illite (or muscovite) can be tolerated as an impurity, but although illite has a higher alumina : silica ratio than kaolinite, its potassium content reduces its thermal stability (Figure 8.16 and associated discussion). Silica contents in excess of those appropriate for kaolinite and mullite reflect the presence of quartz, and the presence of titanium oxide minerals is responsible for the reported TiO_2 contents. The reported K_2O contents relate to the presence of illite.

9.4.2 Sillimanite and 'mullite ore'

There are three aluminium silicate polymorphs, namely sillimanite, andalusite and kyanite, which share the same simple chemical formula (Al_2SiO_5 or AS). They differ in their crystallographic structures and stabilities (with respect to both pressure and temperature), and occur naturally as components of metamorphic rocks. All three are produced as raw materials, but they are not always clearly distinguished from one another in production statistics. Of the three, sillimanite is most desirable as a refractory raw material, as it fires with little or no change in volume and requires no prefiring (unlike kyanite, which expands by 18% when calcined: this is done to prepare the mineral before use).

When sillimanite (and the other aluminium silicate polymorphs) are fired, they decompose to yield a mixture of mullite ($Al_6Si_2O_{13}$) and a silica mineral:

$$3Al_2SiO_5 \rightarrow Al_6Si_2O_{13} + SiO_2 \qquad (9.2)$$

Mullite is an extremely stable refractory mineral, which does not occur naturally in mineable quantities. Sillimanite is therefore used as a 'mullite ore', providing an adequately pure mullite-based refractory, suitable for use in glass tanks and in molten metal containment. For very demanding applications, pure mullite refractories may be required, which can be produced by firing bauxite with kaolinite.

9.5 OTHER REFRACTORY PRODUCTS AND RAW MATERIALS

9.5.1 Zirconia and zircon

Zirconia, ZrO_2, is a highly refractory oxide prepared by firing zircon ($ZrSiO_4$) with dolomite:

$$ZrSiO_4 + CaMg(CO_3)_2 = ZrO_2 + CaMgSiO_4 + 2CO_2 \qquad (9.3)$$

OTHER REFRACTORY PRODUCTS AND RAW MATERIALS

Zirconia melts at temperatures in excess of 2200°C, but is generally used as a component of zirconia–aluminosilicate (AZS) refractories, where it enhances stability. A typical AZS refractory might contain up to 20% ZrO_2, with 60% Al_2O_3 and 20% silica. AZS refractories are widely used for applications involving contact with molten glasses.

In addition to being a source of zirconia, the mineral zircon is used in its own right as a refractory sand. Most of the world's zircon is produced from beach sands, with Australia dominating production. Although expensive, zirconia sand is used for particular applications (such as moulding and also for lining ladles and nozzles in steel making) where its durability in service or lack of reaction with the molten metal compensates for the increased cost.

9.5.2 Olivine

Olivine, particularly forsterite (Mg_2SiO_4), is used as a refractory for certain applications (Griffiths, 1989). It used to be used in electric night storage heaters, in view of its high thermal capacity, but has been superseded in this application by magnetite. It is otherwise used as an alternative to quartz sand as a facing material in casting and founding processes – it has a high melting point (1760°C), and a low uniform thermal expansivity. However, because of its higher cost (four times that of quartz sand) olivine 'sand' is only used in metal casting where it is necessary to prevent reaction between molten metal and sand. For example, austenitic manganese steel (14% Mn) reacts with silica sand, which melts and bonds with the metal; this problem is avoided if olivine 'sand' is used instead.

Sources of olivine consist essentially of dunites, which are quarried, crushed and sized to give powders with the required grain size. World production is dominated by Norway, with other producers in Austria and Spain, and in the United States (Washington State and North Carolina). Material which is serpentinized requires calcining prior to sale, and may not be suitable for certain applications. In general, the market for olivine is somewhat at a low ebb at present, but new, non-refractory, markets such as sand blasting (to overcome the health hazard of silicosis associated with quartz sand) are becoming increasingly important.

9.5.3 Chromite

Chromite is a valuable refractory raw material, often used in combination with magnesia or other materials to produce particular refractories (Mc-Michael, 1989). Ideally, chromite has the chemical formula $FeCr_2O_4$, but as a spinel it shows compositional variation, principally with substitution of Mg for Fe and Al for Cr. For refractory purposes, the silica content of a chromite ore should be below 3.5%.

In addition to use with magnesia, chromite on its own is used like olivine as a foundry sand. It is more expensive again, but is similarly unreactive for steel casting and has better thermal expansivity properties.

Chromite is derived from ultrabasic rocks, mainly from the layered igneous rocks of the Bushveld complex in South Africa. Podiform chromites (in which the chromite is 'pelletized' within a serpentinized matrix) provide an alternative source, with major deposits located in Kazakhstan, Albania, Greece, Turkey and the Philippines. There are no active sources in the UK, although it does occur in ultrabasic rocks on Rhum and Skye, and has been worked in the past in the Shetlands.

9.5.4 Diatomite (diatomaceous earth or kieselguhr)

Although not strictly refractories in their own right, bricks produced from diatomite are used in many foundry applications to provide thermal insulation between high and low temperature parts of an assembly. Diatomite bricks are distinctly low in density, making them ideal for roof construction: with a specific gravity of 0.65 they will float on water.

Diatomite is a sediment composed predominantly of the siliceous tests of microorganisms; mineralogically, it consists of opalline silica. It typically contains 86–94% SiO_2, with significant quantities of alumina (derived from clay impurities) and small amounts of iron oxide. Diatomaceous earths are produced from both marine and freshwater lake deposits: in both settings, input of land-derived sediment is minimal, and an association with volcanic sediments in many cases suggests that there may be a volcanic source for the silica. Diatomite has many uses other than refractory manufacture, which only accounts for a small proportion of production (Harben and Bates, 1990).

9.6 APPLICATIONS OF REFRACTORIES

From the foregoing discussion of the different refractory products, there appears to be an extremely wide range of materials which are available. The designer of a kiln, furnace or metallurgical process will specify particular refractories for particular purposes. An example is shown in Figure 9.4, which illustrates the use of different types of refractory bricks in the lining of a rotary kiln used for cement manufacture. The calcining zone is lined with aluminosilicate bricks (produced from chamotte), chosen and arranged so that their alumina content increases towards the burning zone. The burning zone is lined with magnesia–chrome bricks, and the cooling zone contains bricks with the highest alumina content, produced from bauxite-bearing raw material blends. This zonation reduces the likeli-

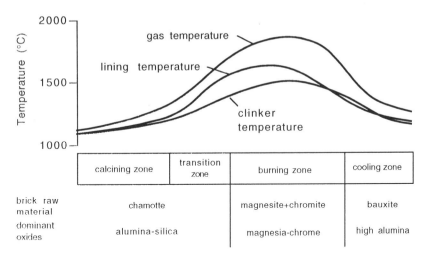

Figure 9.4 Distribution and composition of refractory bricks used to line a rotary cement kiln.

hood of reaction between the cement clinker and its containment, and serves to ensure that the clinker composition is maintained within the C_2S–C_3S–C_3A triangle at all stages (cf. Chapter 7). Rather more complex arrangements of different brick types are used in the glass industry, with zirconia–aluminosilicate (AZS) bricks used in glass containment and the widespread use of high purity mullite or other aluminosilicate bricks. It is essential that the refractory bricks which come into contact with glass are low in iron, to prevent contamination of the glass batch (cf. Chapter 6).

Glass and cement manufacture require that reaction between the refractories and the contained melt or clinker is minimized, and a similar requirement holds for iron and steel production, in which slags form an essential part of the metal refining process. However, it is beyond the scope of this book to deal in detail with the mineralogy and chemistry of slags, which is a matter of immense importance to metallurgists. In iron and steel production, the raw materials used to charge a blast furnace include iron ore (magnetite), coke and limestone, and a number of reactions take place, under strongly reducing conditions, to yield a calcium silicate slag and metallic iron, which can be drawn off as liquids, and carbon dioxide gas. The formation of the slag helps to remove impurities such as phosphorus and sulphur which are present within the iron ore, and to prevent the formation of metallic Si within the strongly reducing conditions of the blast furnace. On cooling, slags form glasses with varying colours, often in shades of green, brown and blue because of the presence of iron compounds. Once they crystallize, slags contain a suite of minerals characteristic of silica-undersaturated basic igneous rocks and cement

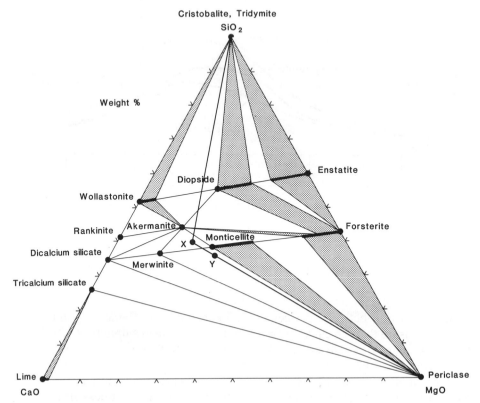

Figure 9.5 Phase relations for the system SiO_2–MgO–CaO (weight %), showing slag compositions X and Y, whose reactions with silica and magnesia refractories are discussed in the text. The compatibility triangles described in the text are shaded for clarity.

clinker (Chapter 7), often showing attractive skeletal crystal shapes as a consequence of rapid crystallization (Figure 2.8f).

Slag compositions are typically dominated by lime, iron and magnesium, and so are strongly silica-undersaturated. Magnesia-based bricks are generally used in their containment, but react in service and require replacement. The life of a magnesia refractory can be extended, however, by conditioning the slag to increase the compatibility of the slag and the refractories. The way in which this can be done is illustrated in principle with reference to the phase relationships for the system SiO_2–MgO–CaO (Figure 9.5). In this figure, the composition of a hypothetical slag, X, lies within the akermanite–monticellite–merwinite compatibility triangle, which represents the solid phases formed if crystallization was complete. The nature of reactions with either a silica brick or a magnesia brick can be described qualitatively by drawing a line between the slag composition and the SiO$_2$ or MgO apices, and inspecting the phase changes which take place along the line.

If a silica brick is placed in contact with slag X at a nominal 1300°C, the sequence of phases produced by reaction potentially will be:

- silica (tridymite)
- silica + wollastonite + diopside
- wollastonite + diopside
- wollastonite + diopside + akermanite
- diopside + akermanite
- diopside + akermanite + forsterite
- akermanite + forsterite
- akermanite + forsterite + monticellite
- akermanite + monticellite
- akermanite + monticellite + merwinite.

Note that the phase assemblage changes by one phase each time a tie line is crossed, and that only two phases coexist on each tie line. In practice, an idealized sequence such as this would be represented by a series of reaction rims surrounding cores of unreacted silica, and would be incomplete.

If a magnesia brick is similarly placed in contact with the same slag, the sequence of reactions is limited to:

- magnesia (periclase)
- magnesia + monticellite + merwinite
- monticellite + merwinite
- akermanite + monticellite + merwinite.

If the slag is conditioned by addition of 10% MgO (e.g. as magnesite), it now lies within the three-phase triangle magnesia + monticellite + merwinite (Y), and so magnesia is a stable phase. In such a case a magnesia brick will be more stable chemically, although it will still react to give monticellite and merwinite. The irregular geometry of contact between the brick and the slag, as a consequence of the slag penetrating the pores within the brick, complicates the idealized reactions described above. In any case, they do not take into account the role of iron, the presence of which strongly influences the nature of the reaction products.

10 Assessment of mineral deposits

The availability of raw materials – in sufficient quantity, of adequate quality, and at an acceptable cost – is of the utmost importance to any industrial minerals operation, as indeed it is to any mining venture. The assessment of quantity and quality of geological materials is a far from trivial subject, and no detailed discussion of the methods which are used will be made here. For information about geostatistics reference should be made to standard texts such as Isaaks and Srivastava (1989). Instead, the purpose of this chapter is to discuss ways in which computers are used in raw material reserve evaluation. There are a number of different software packages available for mineral reserve evaluation, ranging from the relatively unsophisticated Macintosh-based MacOreReserve (which is valuable in circumstances where the geology is very simple, as well as being an easily accessible teaching tool) through to very sophisticated packages such as Datamine, which can handle very large quantities of data and are ideal for complex geology or for situations where for each sample there may be several tens of quality measurements. It is not intended here to critically appraise different software packages, and conversely use of a single package as the basis for discussion does not imply unconditional endorsement of that package. For examples of a number of different applications reference should be made to Annels (1992), bearing in mind that many improvements in practice have been made since that book was published. A reader with serious interests in acquiring mineral reserve evaluation software is recommended to consult a number of suppliers and to choose on the basis of the particular requirements that he or she may have. Irrespective of the different software packages that are available, the treatment of data used for the evaluation is essentially the same.

The basis of this discussion will be a genuine but anonymous kaolin clay deposit, fictitiously set in Cornwall (n.b. no bedded clay deposits are worked in Cornwall) that is currently being mined and that has been thoroughly evaluated using computer techniques. The study is based upon

DATA GATHERING

approximately 100 drillholes intersecting two separate beds of kaolin. Each intersection has resulted in several vertical samples of clay, each of which has been separately analysed in the laboratory for six qualities. Those qualities are, separately and jointly, critical in the selection of raw material to be supplied to the production plant. The qualities are brightness (two different measurements), viscosity, flow rate testing, grittiness, and the level of titanium dioxide.

The computer vehicle used for reserve evaluation and mine planning is a software system called PC/Cores, which is used widely in North America, the United Kingdom and many other countries, and which operates on an IBM-compatible personal computer (or on a high-powered Macintosh with a PC emulator). A more detailed discussion of the computer hardware and software configuration is given in Appendix C.

10.1 DATA GATHERING

The most difficult work associated with computerized reserve evaluation takes place before one approaches the computer. The first task is to collect all (or as much as possible) of the input data that will be required for the study. The data fall into four categories, each of which will be discussed separately: (1) drilling data, which include, for each drillhole, the X–Y coordinates and the surface elevation of the drillhole, accompanied by drilling log data which specify the depth and intersection of each geological or lithological horizon encountered; (2) laboratory analysis data, which will be in the form of reports on the qualities of each sample taken from each drillhole; (3) surface topography data, for the calculation of a digital terrain model that will be used in overburden calculations and (4) property and planning maps, which are necessary for limiting the reserve calculations and describing the mine plan. An example of a typical borehole report, of the type generated by PC/Cores, is shown in Figure 10.1.

10.1.1 Drilling log data

The drillhole log contains data about the location and structure of a drillhole. The following elements of data are included:

- an X-coordinate and a Y-coordinate marking the geographic location of the drillhole in terms of easting and northing according to some acceptable cartesian coordinate system.
- the surface (or 'collar') elevation of the drillhole.
- a unique identifier for the drillhole. A common form of hole 'ID' is the year in which it was drilled, followed by a drilling sequence number, for example, '93–14'.

UOM Geology Computer Study

North	8800	East	12000
File	dh010	ID: 93-10	

Drilled on 16 July 1994
Drilled by Reliable Drilling Ltd
Laboratory Analysis by Core Analysers
Computer Data Processing by Mentor
Figure 10 - 1

Lin Num ber	Strata Roof		Thick- ness Metres	Lithological Definitions		Quality Data...					
	Depth	Elev.		Code	Descriptions	Brit1	Brit2	Flow	Viscos	Ti-02	Grit%
1	0.00	145.0	26.00	ovb	Overburden						
2	26.00	119.0	6.00	ks2	Kaolin : Seam 2	86.70	90.40	68.75	255.00	1.08	0.40
3	32.00	113.0	6.00	pt2	Parting within Seam 2						
4	38.00	107.0	2.00	pt2	Parting within Seam 2						
5	40.00	105.0	6.00	ks3	Kaolin : Seam 3	83.90	89.10	72.25	265.00	0.46	1.10
6	46.00	99.0	4.00	ks3	Kaolin : Seam 3	82.80	89.10	72.75	265.00	0.46	0.20
7	50.00	95.0	4.00	ntk	Low Quality - Not Tested						
8	54.00	91.0	3.00	nck	Non-Commercial Kaolin						
9	57.00	88.0	5.00	nck	Non-Commercial Kaolin						
10	62.00	83.0	2.00	ntk	Low Quality - Not Tested						
11	64.00	81.0	10.00	snd	Sand						
12	74.00	71.0	0.00	eoh	End of Drillhole						

Figure 10.1 A typical borehole report showing location, stratigraphy and clay quality parameters.

- optionally, notes or comments about the drillhole, such as the names of the supervising geologist and the driller, and any other comments that could be useful to the future users of the log.

Following those general drillhole data, the actual drilling log data must be available and will include, for each structure or formation encountered during the drilling operation, the following information:

- the depth at which the horizon was encountered (the 'roof')
- the thickness of the formation (or the 'floor' depth)
- a lithology code identifying the material encountered, such as 'clay', 'sand' or 'limestone', etc.
- optionally, any additional descriptive data which might help in future use of the data, such as 'shiny' or 'argillaceous'.

It is necessary to create a list of all of the lithological names that might occur in any of the drillholes, since the computer software will require such a list for drillhole data entry and validation.

10.1.2 Laboratory quality data

Core samples from the drillholes will have been sent to the laboratory for quantitative analysis of the qualities that are of interest. It is important to agree with the laboratory, in advance, on the format in which quality data will be recorded. A well-organized reporting format will ease the task of connecting laboratory data back-up with the right drillhole and the right sample from that borehole.

A core sample from a single seam may be broken down into multiple samples, each representing a vertical subset of that seam, and each possessing its own set of qualities. It is of critical importance that each subsample be correctly identified with the corresponding thickness and depth on the drilling log. Larger mining companies that operate their own laboratories frequently invest heavily in computerized systems for sample identification and tracking as a sample moves through the laboratory. Smaller companies must take extra care to ensure proper control of laboratory quality data.

10.1.3 Surface topography data

If the mining method is to be 'opcncast' or surface mining, then the thickness and volume of overburden becomes a key factor in mine economics. In order to calculate overburden thickness and ratios, the computer software will require access to surface elevation data for the creation of a digital terrain model (DTM).

The traditional form in which surface elevation data are available is as a contour map of the area. With increasing frequency, aerial survey companies are now delivering the surface elevation map (with contours and cadastral data) in the form of a computer-aided design (CAD) system drawing file on a PC diskette. The reserve evaluation software then reads the surface elevation contours from that diskette and converts the contour lines and elevations to the format which it needs for calculating and gridding the DTM.

Another excellent use of the DTM is as a tool for the validation of drillhole locations and elevations. The PC software can compare the recorded elevation of each drillhole with the DTM elevation at the drillhole's X- and Y-coordinate location. Any significant discrepancy between the two usually indicates that the drillhole either has the wrong elevation or there is an error in its X- or Y-coordinate.

If the source of surface elevation data is a contour map then the digitizing tablet (discussed in the computer configuration addendum to this chapter; Appendix C) comes into play. It will be used to convert the map's contour lines to digital format to allow the calculation of the DTM.

10.1.4 Property and planning maps

The computer software will be performing its calculations on a rectangular area superimposed with a uniform square grid. Unfortunately, few real-life mining operations are so neatly rectangular. The fourth major element of input data is a mine planning map containing, at a minimum, the following information:

- a mining property boundary beyond which mining cannot take place and outside which reserves should not be calculated;
- all barriers to mining operations, such as railways, highways, power lines, cemeteries and residential areas;
- property ownership lines within the mining boundary, to permit calculation of leases and royalties if there are separate owners.

The pertinent polygons and lines from this map need to be digitized (using the PC's digitizing tablet) and stored in data files for use by the reserve evaluation system. Later, the same map (or a computerized replication of it) will be necessary to lay out the intended mining plan, area by area, for effective mine planning and modelling. The map used in this example, together with the borehole locations, is shown in Figure 10.2.

10.1.5 Data availability

It would be most convenient if all of the necessary data arrived on one's desk at the same time, but that is not the way it usually happens. Rather,

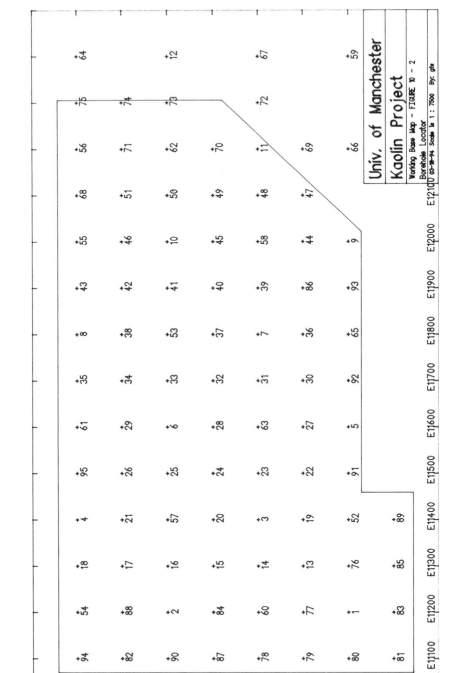

Figure 10.2 Basemap with borehole locations and the property boundary

the data tend to arrive piecemeal at unexpected times. One way or another, when the data are available, they must be entered into the computer system and thoroughly checked, validated and checked again before evaluation and modelling can begin.

10.2 DATA CHECKING AND VALIDATION

There is a tendency, once the last scrap of data has been entered into the computer, to rush ahead with model-building and map-drawing. In fact, this is the moment when the greatest self-discipline is required to ensure that the gathered data are accurate and reliable. There are a number of computer tools that can be of great assistance in this effort, some of which are discussed here.

10.2.1 Borehole profiles

A profiling program permits the concurrent display of multiple boreholes (on the computer screen or on paper) that enables a visual check of the accuracy (reasonableness) of the boreholes' elevations, X–Y locations and stratigraphic depths. In a case where multiple seams of ore are being evaluated, the borehole profile also provides the ideal tool for seam correlation, i.e. the correct identification of a unique seam as it moves across the propery.

In this effort, one should ensure that every borehole has been pictured in at least one horizontal and one vertical cross-section. The borehole profile shown in Figure 10.3 demonstrates the usefulness of this tool.

10.2.2 Borehole posting map

A basemap showing the location of all of the boreholes (shown in Figure 10.4) is useful to ensure that the X- and Y-coordinates of each borehole have been captured and entered correctly (perhaps the most common data error is the transposition of X- and Y-coordinates). The posting map also serves as a check that all boreholes have been entered into the computer. One may also display one or more data elements on the same map.

10.2.3 Validation via contour maps

Undoubtedly the most effective method of checking the validity of numeric data extracted from borehole files is the **contour map**. The computer software permits the extraction of an X–Y–Z data set containing, for each borehole, the borehole identifier, its X- and Y-coordinates and a Z-value. The Z-value in a given data set may be structural, such as the elevation or

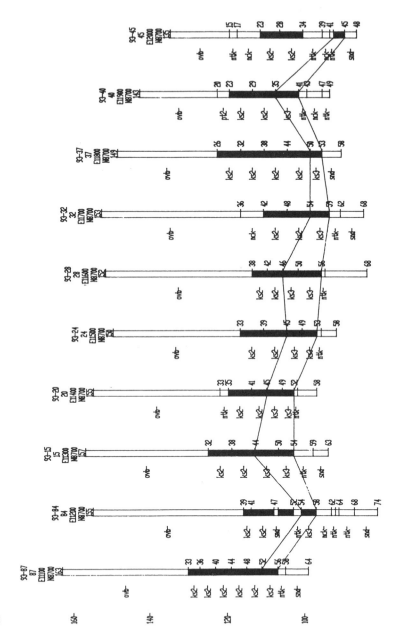

Figure 10.3 Display of 10 boreholes lying along an east–west cross-section of the property, which can be used for seam correlation and to check errors in data input. The lithographical codes are as follows: oub = overburden; Ks2 = Kaolin seam 2; Ks3 = Kaolin seam 3; pt2 = parting within seam 2; ntk = low quality, not tested; nck = non-commercial kaolin; snd = sand.

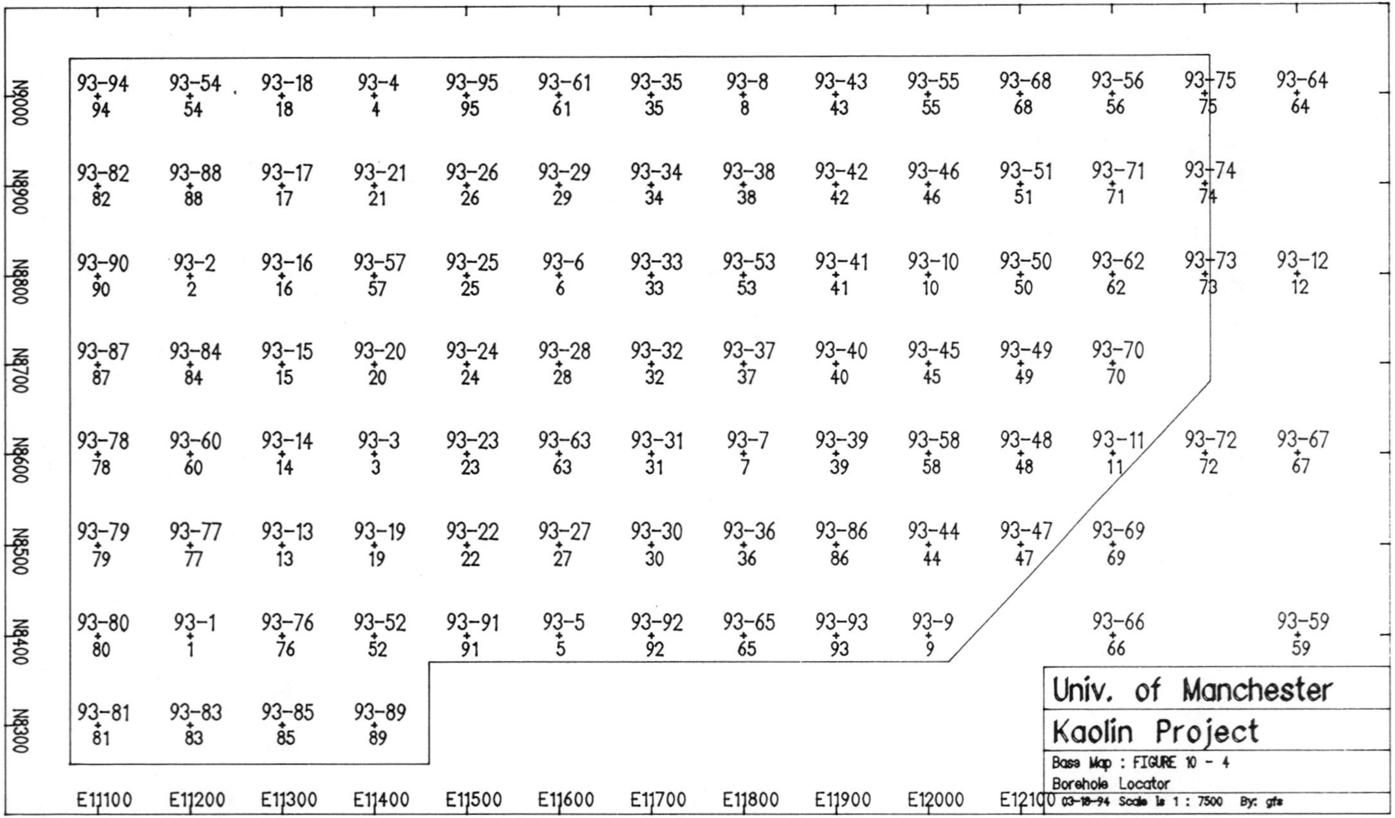

Figure 10.4 Basemap showing borehole locations (crosses); borehole identification codes (above the crosses), PC/Cores files numbers (below the crosses) and the property boundary.

thickness of a stratigraphic unit, or it may be an element of quality data, such as brightness or viscosity.

The software is used to convert the X–Y–Z dataset into a uniform square grid which can then be 'contoured' on screen or on paper. Any significant errors in borehole location or in structural or quality data will be highlighted by a visible discontinuity in the contour map. The borehole from which the offending data was extracted can then be located, the error corrected, and processing continued.

A contour map of total clay thickness (Figure 10.5) shows a 'bull's-eye' type of error (i.e. a series of concentric circles around a too-large thickness value), where the borehole is easily identified and the error promptly corrected. A contour map using the corrected data is shown in Figure 10.6.

10.3 DATABASE DESIGN AND CREATION

In our software example, all of the data for each borehole were entered into a single computer data file for that borehole, with the result that there were as many borehole data files as there were boreholes. This can be an unwieldy form of data organization, and can result in inefficient computer operation. Once one is confident that all of the borehole data files are correct, then the borehole data are collected and compressed into a single computer file called the **borehole database**. Each software system has its own methods of structuring its database, but the differences are not particularly important. What matters is the existence of a complete database, from which any and all data can be accessed, extracted and used at any time. Once the borehole data are stored in a single database (a computer file), the individual borehole data files may be archived to a back-up tape or diskette, freeing up the limited disk storage space on one's computer. It is important always to maintain at least two magnetic copies of your borehole data files: they are expensive and time-consuming to re-enter.

Once the borehole database has been created, the user may commence actual reserve evaluation and mine planning.

10.4 GRIDDING THE DATA

Most computerized geological modelling programs employ the technique of uniform square gridding as an intermediate step in data analysis. The first step in this process is the selection of grid intervals, or grid cell size. First, of course, one must establish a 'grid area', i.e. a rectangle which encompasses all of one's data and all of one's mining property. Then one

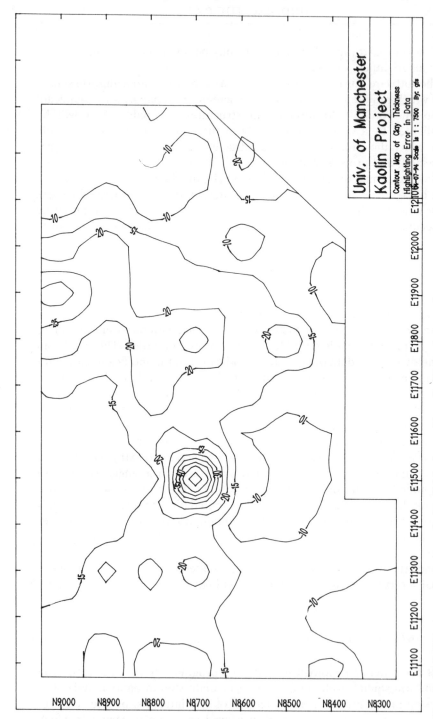

Figure 10.5 Error-revealing map of clay thickness, with a 'bull's-eye' around an erroneous piece of data at about E11525/N8700 (borehole 93–24).

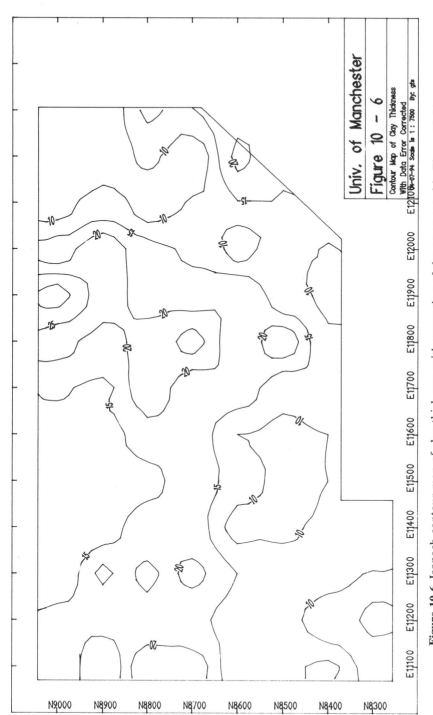

Figure 10.6 Isopach contour map of clay thickness, with correction of the error noted in Figure 10.5.

needs to specify a grid interval so that the grid area can be divided up into uniform square grid cells.

10.4.1 Selection of grid cell size

There is no hard and fast rule for selection of the optimum grid cell size. One approach suggests that the grid interval ought to be about the same distance as the average distance of each borehole from its nearest neighbour. Another school of thought contends that there should be N grid cells for each borehole, where N ranges from 4 to 20. Perhaps the best guideline is that the grid cell size should be what the user thinks it should be. One needs to bear in mind that halving the grid interval has the effect of quadrupling grid and model calculation times. In cases where thickness of the overburden (or the ratio of overburden thickness to ore thickness) is an important economic factor, then the relative steepness of the terrain will become an important factor in the selection of grid interval. (The steeper the terrain, the smaller the grid interval should be to ensure accuracy in overburden calculations.)

In the example problem we have chosen a grid interval of 25 m: the grid pattern is displaced on a basemap in Figure 10.7.

10.4.2 Selection of gridding method

The gridding method employed in the current example problem was 'IDS', the inverse of the distance squared. Specifically, the software follows this procedure:

1. It calculates one grid intersection point at a time
2. At each grid point, it searches for the nearest borehole data value in each of the four quadrants
3. It calculates the distance of each borehole from the current grid intersection point, then uses the inverse function of the square of that distance to 'weight' the Z-value from that borehole.
4. It sums the four weighted values, 'unweights' the sum using the inverse of the distances, and uses that unweighted value as the Z-value for the data set at this grid intersection point.

The end result of the gridding process is a grid file containing a Z-value for each grid intersection point lying inside the property boundary.

10.4.3 Purposes and advantages of gridding

By having a Z-value for each grid point, and by having a grid file for each structural or quality data element that one wants to consider in the reserve evaluation, the software can now proceed with collecting all of the data for

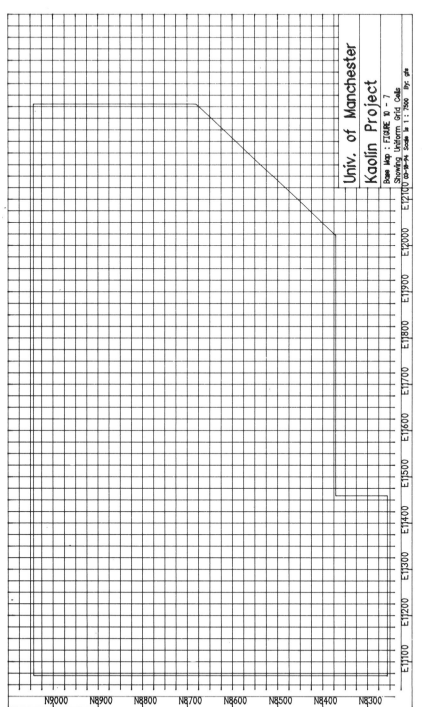

Figure 10.7 Basemap showing the modelling area subdivided into uniform 25 m squares.

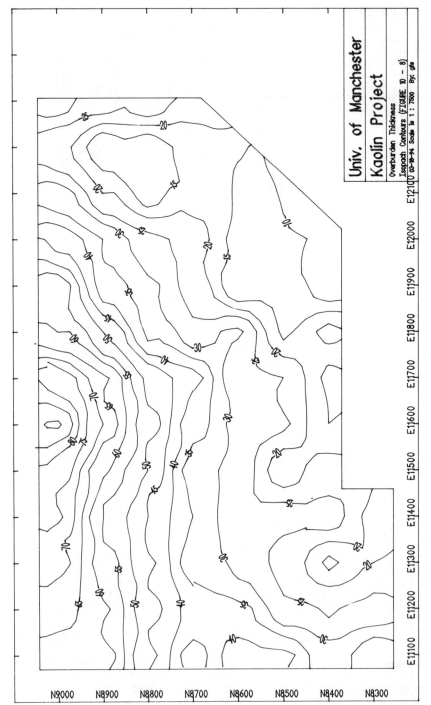

Figure 10.8 Overburden thickness, increasing from south to north.

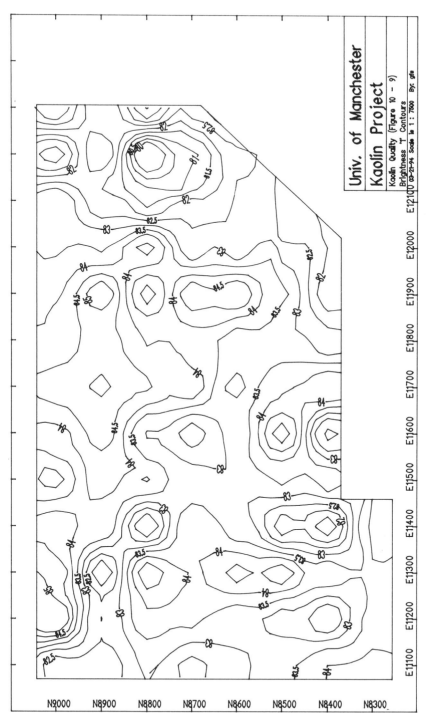

Figure 10.9 Contoured values for quality parameter 'brightness 1'.

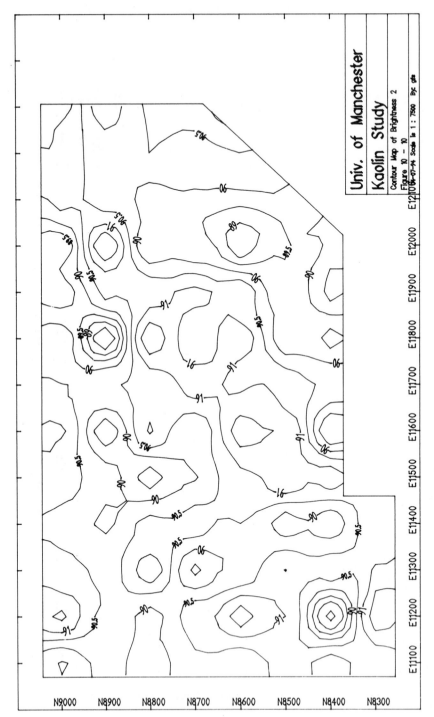

Figure 10.10 Contoured values for quality parameter 'brightness 2'.

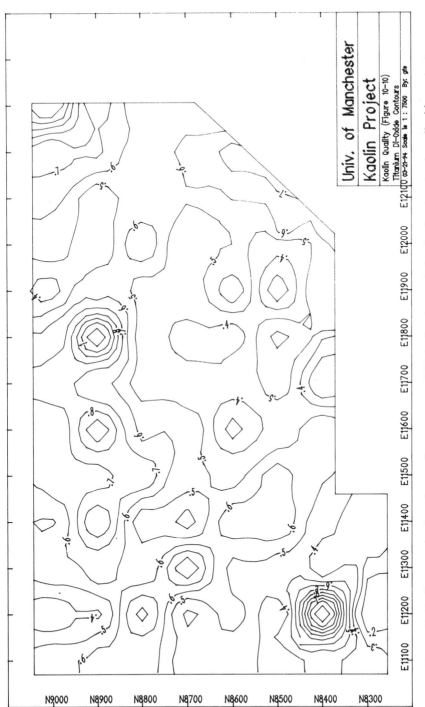

Figure 10.11 Contoured values for quality parameter 'TiO$_2$', showing variation in the clay titanium dioxide content.

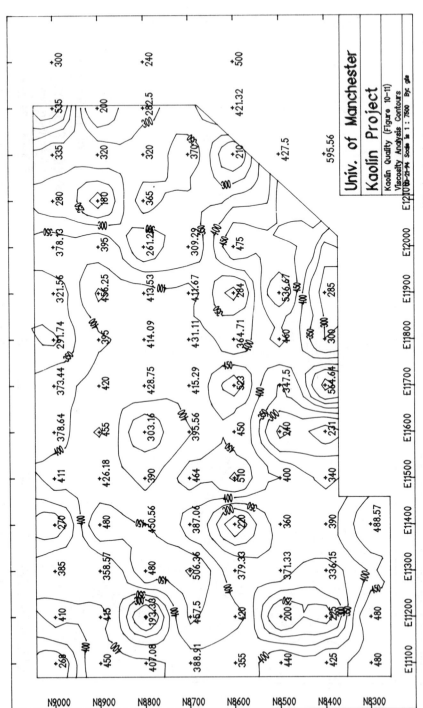

Figure 10.12 Contoured values for quality parameter 'viscosity', showing borehole locations.

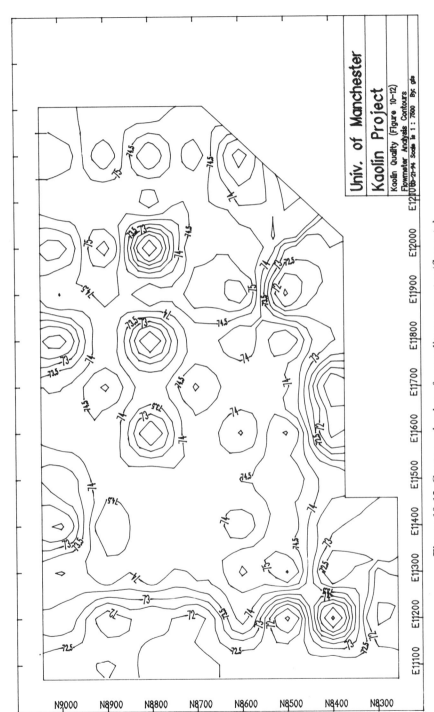

Figure 10.13 Contoured values for quality parameter 'flowrate'.

each grid cell into a single modelling unit. Thus, for the 25 m grid cell centred about the coordinates of East 11700 and North 8700, we will have Z-values for seam thickness, overburden thickness, and each of the six clay qualities which we are modelling: brightness 1, brightness 2, flow rate, viscosity, grit and titanium dioxide. Then, when we define a mine plan which mines all or part of that cell, the computer software will know exactly the volume of clay (25 m × 25 m × thickness), the volume of overburden, and the six qualities of clay contained in that cell.

Another advantage of gridding, of course, is the ability of the software to read a grid file and produce a contour map of its contents. Contour maps have long been, and continue to be, a primary tool in reserve evaluation and mine planning. Contour maps illustrating the spatial variation of the quality parameters included in the model are shown in Figures 10.8–10.13.

10.5 BUILDING THE RESERVE MODEL

Once the required data have been extracted and gridded, another program in the reserve evaluation software system reads the pertinent grids, collects all of the data about each cell into a modelling cell, and stores those data in a model file. In addition to the structural and quality data, the model-builder also stores a cell factor for each cell which represents the percentage of that cell lying inside the property boundary.

Say, for example, that the cell factor at E11500/N8400 is 80%. The software can now calculate the exact volume of clay in that cell that is inside the property boundary:

25 m × 25 m × 12.1 m (thickness) × 0.80 (cell factor) = 6050m^3

Finally, a reserve model need not contain all of the clay within the property. Rather, it can be restricted to containing only the data pertaining to a specific named seam or to a specific selected vertical horizon (or bench). Thus, one can build a model of only that clay lying in a horizontal bench between the elevations of 300 and 310 m above sea level. Separate bench or seam models may then be considered concurrently during the reserve evaluation and mine planning process.

10.6 MODEL EVALUATION AND TESTING

Once the model has been built, one may run the software modelling program to assess the total in-place reserves. Such a report is included here, showing not only total volumes of clay and overburden and weighted averages of each quality, but also a complete set of histograms portraying the distribution of quantitative and qualitative data regarding the deposit.

MODEL EVALUATION AND TESTING

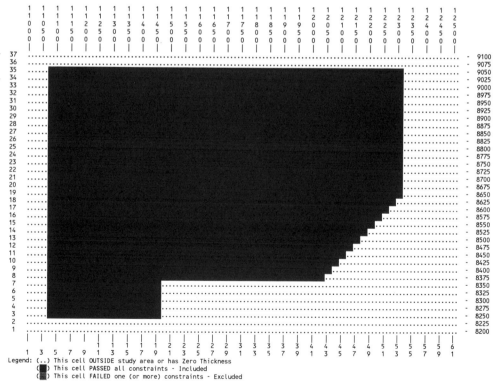

Figure 10.14 Summary report for an unconstrained modelling run (a), including a plan (b) of those cells within the property boundary which pass the constraints (they all do, as this run is unconstrained) and summaries of the quality parameters that are included in the model.

```
03-21-1994                      UOM Geology Computer Study                  16:35:49

MODEL : ALLSEAMS.MDL                                 All Kaolin Seams - WITH CONSTRAINTS

Hectares =      57.18   Volume =  15,035,801 Tonnes   Density =     1.98
Overburden = 13,689,816 CuMtr          0.91 CuMtr per Tonne
Number of cells : In Model = 1414      Accepted : 963      Rejected : 451
```

Constraints and Histogram Splits

Variable	Model Constraints(*)		Histogram Specifications			Weighted Statistics	
	Z-Minimum	Z-Maximum	Band-From	Band-Size	Band-To	Averages	Std.Dev.
Thick	4.00	32.00	3.00	3.00	33.00	13.28	4.67
Burden	6.00	40.00*	0.00	10.00	100.00	23.94	8.53
Brit1	79.25	85.85	79.20	0.80	86.40	83.25	0.87
Brit2	88.50*	91.95	88.00	0.40	92.00	90.48	0.56
Flow	70.68	76.10	70.20	0.60	76.20	73.83	1.04
Viscos	180.04	564.61	160.00	40.00	600.00	384.20	66.26
Ti-O2	0.13	1.32	0.10	0.20	1.40	0.51	0.12
Grit%	0.43	11.20	0.00	2.00	12.00	3.34	1.33

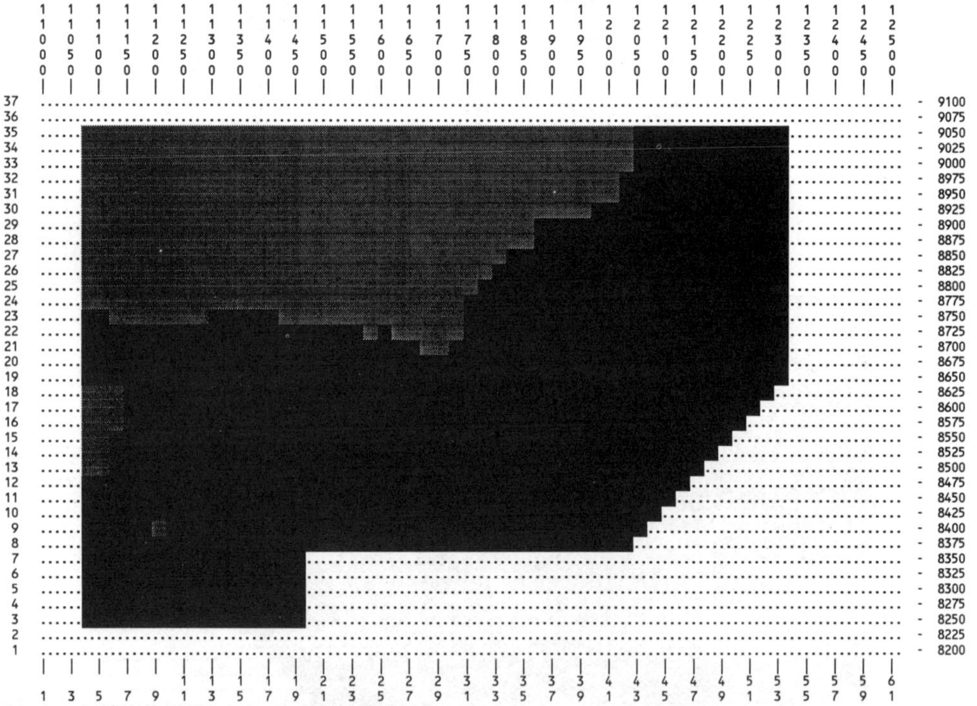

Legend: (...) This cell OUTSIDE study area or has Zero Thickness
        (▒) This cell PASSED all constraints - Included
        (■) This cell FAILED one (or more) constraints - Excluded

Figure 10.15 Summary output for a constrained model evaluation in which the following constraints have been imposed: (1) overburden thickness should be less than 40 m; (2) brightness parameter 'Brit2' must be greater than 88.5. Note that these parameters are identified with an asterisk in the summary model of the model constraints. Note also the distribution of cells which pass those constraints, marked dark on the printer plan of the deposit. This information can be compared with the contour map shown in Figure 10.8, which shows that with the exception of the one cell (E11200/N8400) overburden thickness accounts predominantly for failure to pass the imposed constraints, suggesting that brightness parameter 'Brit2' is not generally influential in this case. Histograms and detailed statistical tables showing variation in quality parameters for this constrained model are given in full in Appendix B.

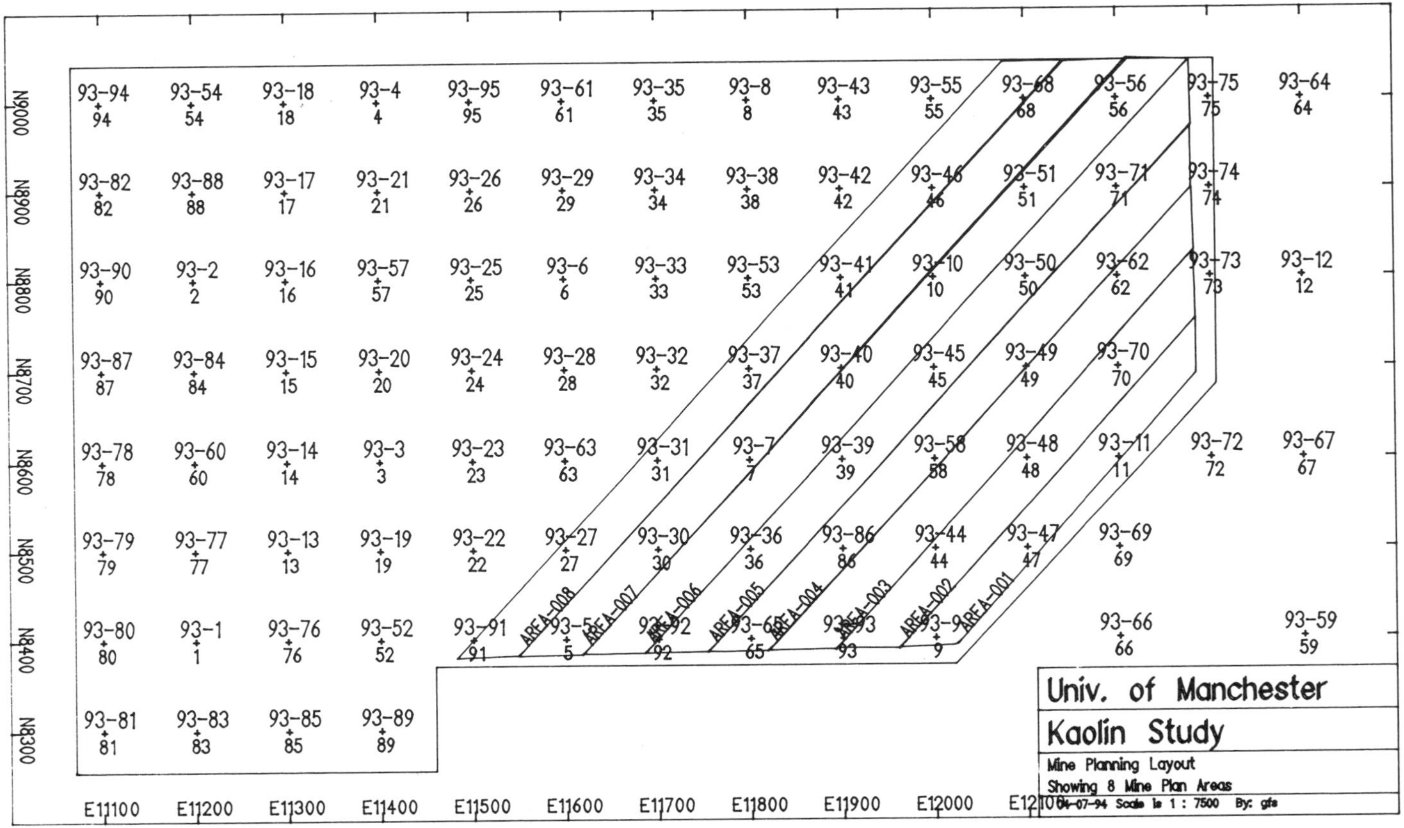

Figure 10.16 Base map showing eight mine planning areas (or 'cuts') to be assessed by the 'Planner' programme.

Area Name : AREA-001.XY

Seam Code	Tonnage	Overburdn	Ratio	Thick	Burden	UOM Geology Brit1	Computer Brit2	Study Flow	Viscos	Ti-02	Grit%	SpGr	03-21-1994 Rec %	17:37:15 Acreage
all	533,609	250,329	0.5	14.27	13.26	82.29	90.22	73.84	349.19	0.64	3.69	2.0	100%	1.89
Tot	533,609	250,329	0.5	14.27	13.26	82.29	90.22	73.84	349.19	0.64	3.69			

Includes 476729 Tons in the Range 0 - 20 Burden

Cumulative Tonnage and Overburden Report

Seam Code	Tonnage	OverBurden	Ratio
all	533,609	250,329	0.47

Area Name : AREA-002.XY

Seam Code	Tonnage	Overburdn	Ratio	Thick	Burden	UOM Geology Brit1	Computer Brit2	Study Flow	Viscos	Ti-02	Grit%	SpGr	03-21-1994 Rec %	17:37:36 Acreage
all	615,597	390,761	0.6	11.31	14.21	82.21	90.01	73.96	366.11	0.57	3.56	2.0	100%	2.75
Tot	615,597	390,761	0.6	11.31	14.21	82.21	90.01	73.96	366.11	0.57	3.56			

Includes 538117 Tons in the Range 0 - 20 Burden

Cumulative Tonnage and Overburden Report

Seam Code	Tonnage	OverBurden	Ratio
all	1,149,206	641,090	0.56

Area Name : AREA-003.XY

Seam Code	Tonnage	Overburdn	Ratio	Thick	Burden	UOM Geology Brit1	Computer Brit2	Study Flow	Viscos	Ti-02	Grit%	SpGr	03-21-1994 Rec %	17:37:44 Acreage
all	669,108	472,075	0.7	10.29	14.37	82.12	89.83	73.88	379.06	0.52	3.60	2.0	100%	3.28
Tot	669,108	472,075	0.7	10.29	14.37	82.12	89.83	73.88	379.06	0.52	3.60			

Includes 595869 Tons in the Range 0 - 20 Burden

Cumulative Tonnage and Overburden Report

Seam Code	Tonnage	OverBurden	Ratio
all	1,818,313	1,113,165	0.61

Area Name : AREA-004.xy

Seam Code	Tonnage	Overburdn	Ratio	Thick	Burden	UOM Geology Brit1	Computer Brit2	Study Flow	Viscos	Ti-02	Grit%	SpGr	03-21-1994 Rec %	17:37:52 Acreage
all	741,324	541,652	0.7	10.71	15.49	82.55	89.74	74.01	380.34	0.49	2.93	2.0	100%	3.50
Tot	741,324	541,652	0.7	10.71	15.49	82.55	89.74	74.01	380.34	0.49	2.93			

Includes 660752 Tons in the Range 0 - 20 Burden

Cumulative Tonnage and Overburden Report

Area Name : AREA-005.xy					UOM Geology Computer Study								03-21-1994	17:38:00
Seam	Tonnage	Overburdn	Ratio	Thick	Burden	Brit1	Brit2	Flow	Viscos	Ti-02	Grit%	SpGr	Rec %	Acreage
all	1,051,284	717,590	0.7	12.54	16.95	83.23	90.06	74.37	373.36	0.52	2.43	2.0	100%	4.23
Tot	1,051,284	717,590	0.7	12.54	16.95	83.23	90.06	74.37	373.36	0.52	2.43			

Cumulative Tonnage and Overburden Report			
Seam Code	Tonnage	OverBurden	Ratio
all	3,610,920	2,372,407	0.66

Includes 840546 Tons in the Range 0 - 20 Burden

Area Name : AREA-006.xy					UOM Geology Computer Study								03-21-1994	17:38:08
Seam	Tonnage	Overburdn	Ratio	Thick	Burden	Brit1	Brit2	Flow	Viscos	Ti-02	Grit%	SpGr	Rec %	Acreage
all	1,246,520	939,264	0.8	13.95	20.81	83.49	90.43	73.99	356.62	0.51	2.36	2.0	100%	4.51
Tot	1,246,520	939,264	0.8	13.95	20.81	83.49	90.43	73.99	356.62	0.51	2.36			

Cumulative Tonnage and Overburden Report			
Seam Code	Tonnage	OverBurden	Ratio
all	4,857,441	3,311,671	0.68

Includes 534060 Tons in the Range 0 - 20 Burden

Area Name : AREA-007.xy					UOM Geology Computer Study								03-21-1994	17:38:17
Seam	Tonnage	Overburdn	Ratio	Thick	Burden	Brit1	Brit2	Flow	Viscos	Ti-02	Grit%	SpGr	Rec %	Acreage
all	1,346,775	1,160,171	0.9	14.91	25.43	83.54	90.72	74.03	355.51	0.49	2.30	2.0	100%	4.56
Tot	1,346,775	1,160,171	0.9	14.91	25.43	83.54	90.72	74.03	355.51	0.49	2.30			

Cumulative Tonnage and Overburden Report			
Seam Code	Tonnage	OverBurden	Ratio
all	6,204,216	4,471,842	0.72

Includes 100613 Tons in the Range 0 - 20 Burden

Area Name : AREA-008.xy					UOM Geology Computer Study								03-21-1994	17:38:26
Seam	Tonnage	Overburdn	Ratio	Thick	Burden	Brit1	Brit2	Flow	Viscos	Ti-02	Grit%	SpGr	Rec %	Acreage
all	1,408,743	1,250,973	0.9	16.44	28.91	83.61	90.91	74.33	393.86	0.46	2.79	2.0	100%	4.33
Tot	1,408,743	1,250,973	0.9	16.44	28.91	83.61	90.91	74.33	393.86	0.46	2.79			

Includes 46694 Tons in the Range 0 - 20 Burden

Figure 10.17 Output from a run of the 'Planner' programme, showing quantities and qualities of the clay in each of the eight mine planning areas shown in Figure 10.15, identified as Area Name: AREA 001.xy–AREA 008.xy.

A complete *in situ* run of the modeller is included in the reports for our example problem (Figure 10.15). Before performing that calculation, one should have in mind one's own best estimates of quantities and qualities, to ensure that the model has been properly constructed and that the results are reasonable.

One may then examine the model in as many different ways as desired, constraining the evaluation to specific areas within the property, or constraining the evaluation to report only on the material that meets specific quality and thickness constraints. In the total *in situ* evaluation, the modelling program reported that we had approximately 23 million tonnes of clay. However that tonnage dropped to approximately 15 million when we asked 'how much clay do we have that has a "brightness 2" rating of 88.5 or better, and that lies under less than 40 m of overburden?' (Figure 10.15). The point is that the model can be evaluated with as many different combinations of constraints as one can imagine, and it can be done in remarkably little time.

10.7 MINE PLANNING USING THE MODELS

The final step in model evaluation is the definition of mine planning areas. This is normally done by marking off logical mining areas on the basemap, then digitizing the turning points of the area polygons and saving the digitized polygons on the computer.

The software can then read each polygon's X–Y coordinates and:

- Calculate the area (in hectares or acres) of the polygon
- Calculate the area of each modelling grid cell that lies wholly or partially within the polygon.
- Report on the quantity and qualities of the clay lying in each seam (or each bench or in total) underneath that polygon.
- Save that polygon's data in a file for subsequent use in establishing a mine-planning sequence and setting periodic mine production targets.

Figure 10.16 is a base map showing eight digitized mine planning areas (or 'cuts' and Figure 10.17 gives the quantity and quality of kaolin underlying each of those cuts.

The targeting segment of the mine planning software package can now be used to set an area by area and bench by bench mining sequence and to establish monthly or quarterly production targets for any period of time up to the total life of the mine.

Disposal of waste by landfill | 11

In any industrial minerals operation that involves surface workings, the problem of what to do with the site after extraction has to be faced (and as far as possible to be solved and properly costed) before submission of a planning application and commencement of work. Existing operations may be undertaken on the basis of permission granted historically with no provision for restoration and defined end use, but it is good practice (at the very least) to plan for an environmentally satisfactory conclusion to extraction. In heavily populated areas, the interests of local residents often require that a site is restored to permit recreational use, for water or other sports, or for nature conservation. Similarly, in rural areas restoration to previous agricultural or forestry use may be a requirement of planning permission.

It is inappropriate to consider all options for end use here, as they will be very much site-specific. Matters which need to be taken into account generally in consideration of the environmental management of surface workings are discussed at length in government guidelines (Department of the Environment, 1991b), and are essentially the requirements of being a good neighbour. They include the following topics and particular aspects which need to be considered:

- road traffic – size of vehicles on public roads and number of vehicle movements; damage to roads and verges; dirt spilled on roads; visual intrusion, noise, vibration and dust
- blasting – noise, vibration, flying rock, dust
- noise – noise of the mining operation; general machinery noise, warning signals etc.
- air pollution – dust from mining which may be deposited on nearby vegetation; fumes, including emissions from operations carried out in associated use of mined products (fluorine and sulphur dioxide from brick firing, cement manufacture etc.)
- landscape – destruction of skylines by quarrying or creation of an unnatural landscape by tipping

- groundwater – mining operations may affect the quality and quantity of water available for aquifer recharge; dewatering may change the flow of water to pre-existing springs and wells, cause settlement or change water quality; effluents from mining may cause chemical and physical (e.g. suspended solids) contamination of aquifers
- surface water – mining may change flow patterns, drainage and may cause chemical and physical contamination of surface waters (as well as aquifers)
- waste – mine waste includes overburden, rock which is below grade or otherwise rejected, and processing residues, which may be contaminated chemically. Mine waste may be backfilled within the site or removed to an alternative site
- public rights of way – disturbance of roads, footpaths or other pre-existing access
- effects on local residents – an operator must communicate and discuss intentions and plans with local people, and take advice on local factors.
- effects on others – tourists, visitors and investors may be affected by the presence of a mining operation
- heritage – conservation of archaeological sites, cemeteries, sites of special scientific interest, animal and plant habitats.

Many of these matters arise from consideration of the interests and needs of other people who may not benefit directly (or perceive any benefit) from the planned mineral workings. Local conditions will stipulate guidelines to restrict and monitor factors such as vehicle movements, emissions of dust and gases and effects on water.

11.1 USE OF EXHAUSTED MINERAL WORKINGS FOR WASTE DISPOSAL BY LANDFILL

A particular use for exhausted quarries and pits is as a site for landfill (sanitary) waste disposal. The need for such sites to be close to centres of population which generate waste is entirely consistent with the original need for the site of mineral extraction to be close to its markets, and planning conditions are again generally consistent for both extraction and landfill operations. However, landfill disposal of waste is in itself a highly regulated operation where technical details play an important role in site planning: as pointed out by Daniel (1993b), waste disposal technology (for containment of waste) is driven by regulation, not by engineering factors. From the point of view of an industrial minerals operator, the possibility of using the site for landfill waste disposal provides two important possible benefits: (1) the generation of an additional stream of income from the sale of excavated space (£5 per cubic metre in 1993; Harries-Rees, 1993), and

(2) savings in energy costs in associated mineral processing or manufacturing processes which arise from the use of landfill gas either directly (by combustion) or indirectly (when it is used for electricity generation). From the point of view of an industrial minerals supplier, the regulations which govern the design of landfill sites may require preparation of the site by installation of a liner, which may be a layer of bentonite clay or a bentonite–geotextile composite mat. This provides an important market for producers of industrial clays.

In most industrial countries, landfill practice is tightly controlled by regulations imposed by national governments. In the United States, the Resource Conservation and Recovery Act and the Hazard and Solid Waste Amendments to this act underlie all waste disposal planning activities (Daniel, 1993a). In Britain, regulatory guidance is provided by the Waste Management papers of the Department of the Environment (Department of the Environment, 1986; 1989). These regulations (and similar ones in other countries) affect both the classification of waste into different categories, according to perceived hazards, and the design and licensing of landfill sites to receive particular types of waste. The following discussion is based on British practice, but the principles and scientific aspects of the degradation of waste are generally applicable elsewhere.

Waste can be divided into three categories – household, industrial or commercial. Household waste is derived from domestic or residential premises, industrial waste from manufacturing or processing plants and commercial waste from service industries and offices (business, commerce, recreation etc.). Special arrangements are made for 'special' wastes, defined as substances which are dangerous to life (particularly chemicals), which have low flash points (below 21°C; waste fuels and solvents) or which are prescribed medicines.

11.2 COMPOSITION OF WASTE

The composition of municipal waste, dominated by household refuse, divides into organic and inorganic components (Figure 11.1). By weight, the organic fraction makes up about 60% of municipal waste, and consists of plastics (about 5%), paper (about 30%) and putrescible material (garden/yard and food wastes, 25%). It is this putrescible component which is the most important factor in determining the way in which municipal waste changes once it has been deposited; plastics and paper (predominantly cellulose) are comparatively inert.

Putrescible waste consists largely of discarded raw or cooked vegetable or animal matter from the kitchen or dining table, or the garden. It is an ideal source of food for bacteria, fungi and other lowly animals (including vermin), and within a landfill waste disposal site the decomposition of

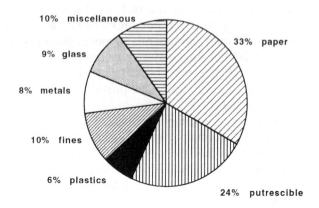

Figure 11.1 Typical proportions (on the basis of weight) of the components of municipal waste in (a) the UK (Department of the Environment 1986; Blakey and Maris, 1990) and (b) the US (Daniel, 1993a).

putrescible waste as a consequence of microbiological activity is an essential and important process. The composition of food is described in terms of its content of proteins, lipids (fats) and carbohydrates, which predominate, and minerals, vitamins etc. which are present in minor quantities (see Coultate, 1992, for a very accessible description of food chemistry). Proteins, lipids and carbohydrates decompose according to the general scheme shown in Figure 11.2, which essentially involves a sequence of microbiological reactions involving distinct populations of bacteria, and which may overlap. It is most important to note that this degradation takes

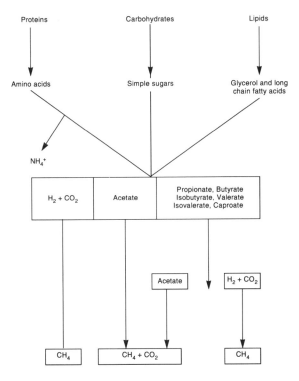

Figure 11.2 Decomposition pathways for the major components of putrescible waste (from Department of the Environment, 1989).

place under anaerobic conditions, which prevail within a matter of days after the waste has been isolated from the atmosphere. The consequences of degradation are that the solid waste is stabilized, which is highly desirable, and that potentially hazardous liquids (leachates) and gases (landfill gas) are evolved and require management.

11.3 LANDFILL LEACHATE

Landfill leachate is never forgotten once encountered. It is chemically extremely complex, and shows systematic variation in composition, depending on the age of the waste from which it is derived and the stage that degradation has reached. Once anaerobic conditions have been established, leachates are characterized by high solute contents and high volatile fatty acid (VFA) contents (Table 11.1 and Figure 11.3). They are frequently black in colour, cloudy or opaque because of the presence of suspended or colloidal matter, and have an unpleasantly obnoxious smell.

Table 11.1 Typical composition of leachates from recent and aged domestic waste (Department of the Environment, 1986). All values in mg/L, except for pH value

Component	Leachate from recent wastes	Leachate from aged wastes
pH value	6.2	7.5
COD (Chemical Oxygen Demand)	23800	1160
BOD (Biochemical Oxygen Demand)	11900	260
TOC (Total Organic Carbon)	8000	465
Total Fatty Acids (as C)	5688	5
Ammoniacal nitrogen	790	370
Oxidized nitrogen (nitrate)	3	1
Phosphate	0.73	1.4
Chloride	1315	2080
Sodium	960	1300
Magnesium	252	185
Potassium	780	590
Calcium	1820	250
Manganese	27	2.1
Iron	540	23
Nickel	0.6	0.1
Copper	0.12	0.3
Zinc	21.5	0.4
Lead	8.4	0.14

As the leachate evolves, VFA and total solute contents decline, the appearance changes to give translucent or clear pale brown liquids, and the smell (although still obnoxious) sweetens.

Because of the complexity of their composition, it is not usual to undertake routinely a detailed chemical analysis of leachates. Periodic monitoring (monthly) of electrical conductivity (reflecting the solute content), chemical oxygen demand (COD; reflecting the organic acid content and presence of other reduced species), ammonia, pH and temperature is adequate for general management purposes (Table 11.1). Temperatures of leachates (and the decomposing waste from which they are derived) are commonly in excess of 25°C (and have been reported to reach 60°C in some cases) during the initial VFA-rich stage, declining to background temperatures (around 10°C in Britain) as the site matures.

11.4 LANDFILL GAS

Landfill gas is composed typically of methane (65%) and carbon dioxide (35%; Table 11.2), with minor amounts of oxygen and nitrogen derived from the atmosphere and other gases derived from waste solvents or other

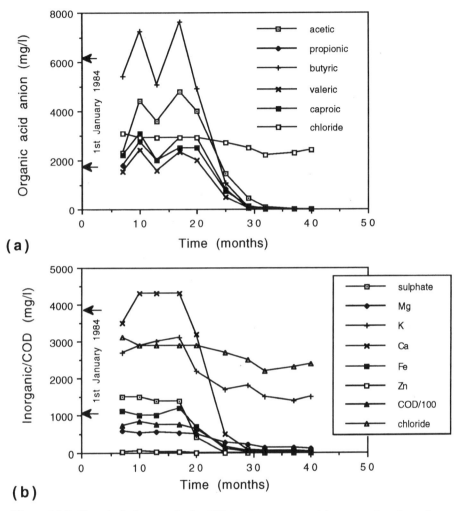

Figure 11.3 Chemical changes in landfill leachate composition as a function of time: (a) organic acids (n-isomers) and (b) inorganics and COD (data from Robinson, 1990). Note the clear transition from young, acetogenic, leachates to more mature, methanogenic leachates.

sources within the refuse. It is generated by microbiological processes, nourished by the organic species dissolved within the leachate. Thus during the initial decomposition of waste the generation of VFAs exceeds the ability of the methanogenic bacteria to generate methane, and so landfill gas is not produced in significant quantities. Landfill gas formation only becomes significant once methanogenic bacteria become established and consume dissolved organic acids at a rate in excess of their production.

Table 11.2 Typical composition of landfill gas (volume %; from Department of the Environment, 1989)

Component	Typical value	Recorded maximum
Methane	83.8	88
Carbon dioxide	33.6	89.3
Oxygen	0.16	20.9[1]
Nitrogen	2.4	87.0[1]
Hydrogen	0.05	21.1
Carbon monoxide	0.001	0.09
Ethane	0.005	0.014
Ethene	0.018	–
Acetaldehyde	0.005	–
Propane	0.002	0.017
Butane isomers	0.003	0.023
Helium	0.00005	–
Higher alkanes	<0.05	0.07
Unsaturated hydrocarbons	0.009	0.048
Halogenated organic compounds	0.00002	0.032
Hydrogen sulphide	0.00002	35.0
Organosulphur compounds	0.00001	0.028
Alcohols	0.00001	0.127
Others	0.00005	0.023

[1] entirely derived from the atmosphere.

The evolution of landfill leachates and generation of landfill gas are coupled processes, and fit within a general scheme of waste decomposition which involves initial acidogenesis followed by acetogenesis and methanogenesis (Figures 11.2 and 11.4). These stages overlap and develop at different times in the development of the landfill, and vary from site to site. Waste management requires the monitoring of both leachate and gas compositions to determine the onset of significant methanogenesis, the containment of leachates to prevent pollution, and the controlled venting of landfill gas to prevent explosions or enable its recovery as a valuable product of the operation. These factors influence the design and operation of a waste disposal site.

11.5 DESIGN OF LANDFILL WASTE DISPOSAL SITES

Although there is a legacy of old landfill waste disposal sites which were designed before the importance of pollution controls was recognized, modern practice is very tightly governed by safety considerations. There is an obvious need to prevent potentially hazardous leachates from entering groundwaters or surface waters close to the site of interest, perhaps reduced if these waters are already unusable because of contamination by,

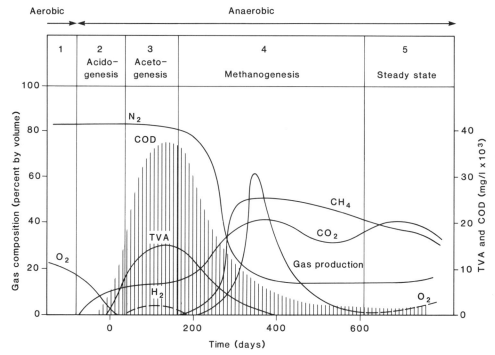

Figure 11.4 Comparison of changes in gas composition and changes in leachate composition (adapted from Department of the Environment, 1986).

for example, salt water intrusion from the sea. However, there is an equal need to dispose of leachates in a safe way, and there are circumstances where this might be achieved by natural processes of attenuation within the aerobic zone of an aquifer. It is also necessary to plan the site to allow management of the gas, particularly if this is to be exploited.

The regulatory authorities in Europe and North America are currently responsible for the licensing of sites to take particular categories of waste, and accordingly require the site to meet certain design specifications. Extensive details of engineering aspects of waste disposal are given in Daniel's book (Daniel, 1993a). There are essentially two types of site: (1) containment sites and (2) attenuate and disperse sites.

11.5.1 Containment sites

A containment site is designed to maximize the retention of leachate within it, and this is achieved by installation of a low permeability clay barrier at the base of the site, or by installation of an artificial liner. The requirements of the regulatory authorities are generally consistent: the clay barrier should be at least 0.9 m (i.e. 3 feet; US Environmental Protection

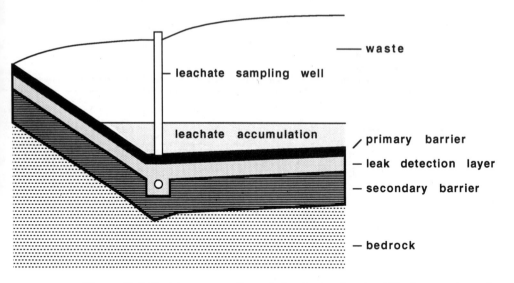

Figure 11.5 Sketch cross-section of the base of a landfill site, to show the arrangement of primary and secondary barriers to leachate migration.

Agency; Daniel, 1993b) or 1.0 m thick (UK practice), and have a hydraulic conductivity of less than 1×10^{-9} m/s. Note that the word 'permeability' refers to a quality of a material, whereas the numerical quantity used to describe this quality is called the 'hydraulic conductivity'. Bentonite clays are most suitable for barrier construction, because of their reactive properties as well as their low permeabilities when compacted. Artificial liners may be impermeable plastics such as high density polyethylene (HDPE) or composite geotextile quilts in which bentonite is sandwiched between synthetic fabrics. HDPE liners are installed in strips welded together on site, and are particularly vulnerable to puncturing or weld failure. It is general practice to install them as part of a composite barrier, in which a secondary barrier underlies the HDPE barrier, with an intermediate drainage layer which can be used to check for leaks (Figure 11.5) and a protective sand layer is placed on top of the plastic membrane. Composite geotextile–bentonite liners are self-sealing, at seams as well as at punctures, but again are often installed as part of a multiple barrier.

11.5.2 Attenuate and disperse sites

Otherwise known as 'dilute and disperse' sites, this type of operation is designed to permit leakage of leachate from the site into underlying rocks, where natural processes clean the leachate. They may be hosted with a moderately permeable rock or sediment, and site preparation may be limited to reworking and compacting the base of the site.

Leakage from sites or across barriers can be estimated using Darcy's law, expressed as:

$$\frac{Q}{A} = ki \quad (11.1)$$

where Q = flow rate (m³ day)
  A = cross-sectional area perpendicular to flow (m²)
  k = hydraulic conductivity (m/day)
  i = hydraulic gradient across the clay membrane (i.e. pressure difference)

For example, i = leachate height/clay thickness, say 3 m/10 m; k is typically 10^{-5} m/day, giving a flow per square metre of 0.3×10^{-5} m³/day.

This shows that the amount of leachate which can pass through a barrier and leave the site is directly proportional to the head of leachate above the barrier, and so an important aspect of site management includes monitoring leachate heads to ensure that limits which form part of the licence are not exceeded. The amount of leachate present within the site can be increased by rainfall, surface or groundwater infiltration (Figure 11.6), or disposal of liquid wastes, and is assessed additionally by consideration of the water balance for the site:

$$L_o = I - E - aW \quad (11.2)$$

where L_o = amount of leachate (m³/y) generated
  I = total liquid input (rain water + infiltrating water + liquid waste; m³/y)
  E = evaporation/transpiration losses (m³/y)
  a = absorptive capacity of the waste (m³/tonne of waste)
  W = weight of waste deposited (tonnes per year).

This approach allows an estimate of the amounts of liquid waste that can safely be disposed of within a landfill site by codisposal (Figure 11.7). In codisposal operations (which generally are not used outside Britain), trenches are cut into the surface of compacted solid waste, to take discharges from tankers which then soak into the solid waste. Many liquid wastes include chemicals which are stabilized or destroyed by reaction with the solid waste, and so codisposal exploits the benefits of the landfill as a bioreactor.

11.6 SITE MANAGEMENT

In the disposal of municiple waste, the requirements of the licence stipulate the way in which the site is run and designed. Typically a landfill waste disposal site will be subdivided into cells, each of which has its own system of leachate collection and monitoring points (Figure 11.8). Waste is

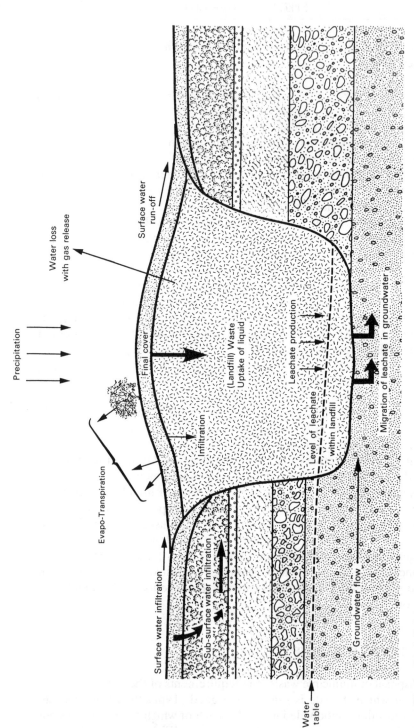

Figure 11.6 Origin of possible sources and losses of water to and from landfill waste (from Department of the Environment, 1986).

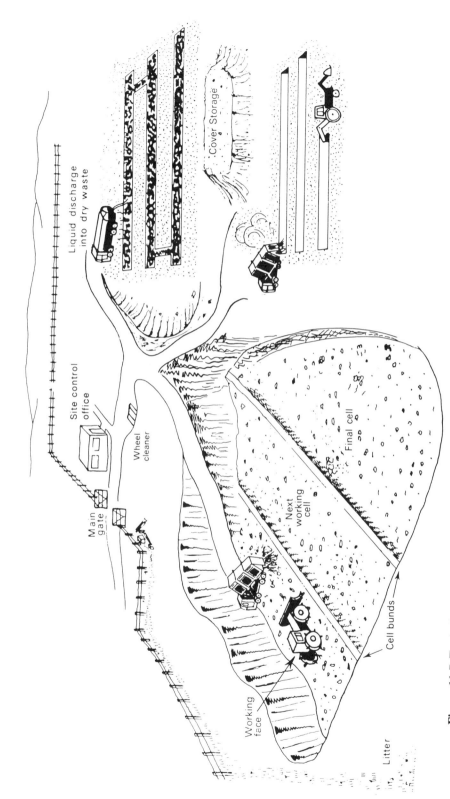

Figure 11.7 Typical arrangements for codisposal of liquid waste (from Department of the Environment, 1986).

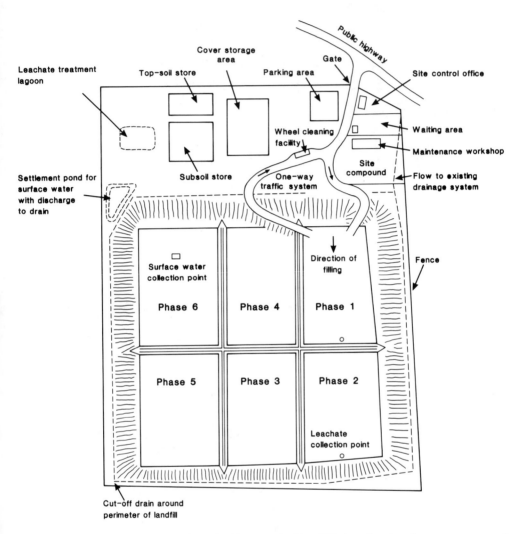

Figure 11.8 Operation plan for a typical landfill site, showing the arrangement of cells (adapted from Department of the Environment, 1986).

deposited rapidly in a big operation, compacted and covered daily by a layer of clay or other material (such as shredded non-metallic automobile waste) to prevent waste blowing from the site or becoming infested by rats and other vermin as well as to reduce rain water ingress. Leachate monitoring points take the form of wells, keeping up with increasing height of the waste by the addition of precast sulphate-resistant concrete rings. Once tipping has been completed the cell is sealed by a cover which may, like the base, be composite, depending on the nature of the site, and a new cell is started (Figure 11.9). On completion of the operation the cover is restored to vegetation by installation of soil and seeding with appropriate crops. Leachate wells are capped with access manholes, and left in an unobtrusive but accessible state.

The nature of these operations results in a heterogenous body within which leachate and gas migration are influenced by the development of local seals and perched water tables (Figure 11.10). There need be no hydraulic connection between cells or even between leachate wells within a single cell. This heterogeneity leads to heterogeneity in the chemical evolution of the leachate–gas system, with certain parts of some cells becoming methanogenic before others. Gas migration can be controlled by the installation of collection systems within the cap, but in the same way that leachate can drain through the base of a site, gas can leak through the cap. Routine, periodic monitoring of soil gas compositions to detect high levels of methane is normal on completion; in extreme cases high soil methane contents are revealed by vegetation which is yellowing and dying back.

In an investigation of the potential of the site to yield economically valuable quantities of methane it is necessary to drill gas wells through the cap, connect them to a manifold and to a flare while pumping trials are undertaken. This is the only way to determine whether or not a site is generating enough methane to be exploited, and the time at which this stage is reached cannot be predicted reliably.

Leaks of leachate from waste disposal sites are monitored by an external ring of boreholes within adjacent rocks or sediments. Within the site, drainage from the cap is collected in a ditch running round the site, which also collects any leachate originating from seeps through the cap (which are unusual but can occur as a consequence of excessive leachate generation in which leachate heads rise higher than the lowest point of the cap).

11.7 SITE GEOLOGY AND NATURAL LINING MATERIALS

In an initial investigation, the bedrock geology of a prospective landfill site is determined, and in many circumstances little or no additional lining material might be necessary. In clay pits used for the brick industry the clay may well be adequate as a base to the site, although care has to be taken to note the presence of heterogeneities such as sand lenses or structural discontinuities which might allow leachates to escape. In particular, if any beds within a face can be seen to be 'weeping' or are persistently damp (demonstrating that water can enter the site using these conduits) there is the potential for water to escape, and so there is a need to seal the site.

Sand, gravel and hard-rock quarries are generally within rocks which are highly permeable either by virtue of interconnecting pores (sand and

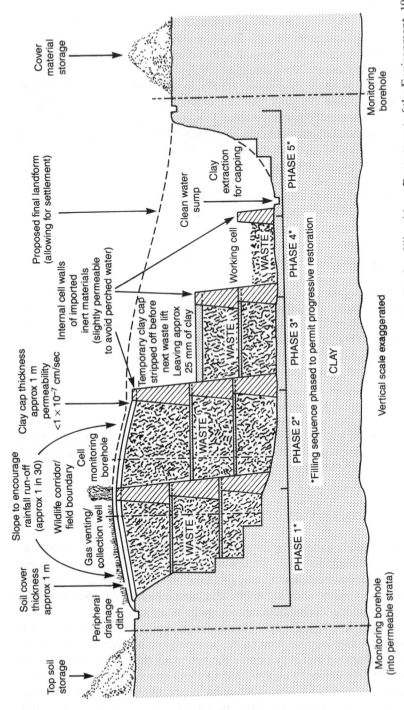

Figure 11.9 Cross-section of active landfill site on clay strata, to show cover of cells after filling (from Department of the Environment, 1986).

SITE GEOLOGY AND NATURAL LINING MATERIALS

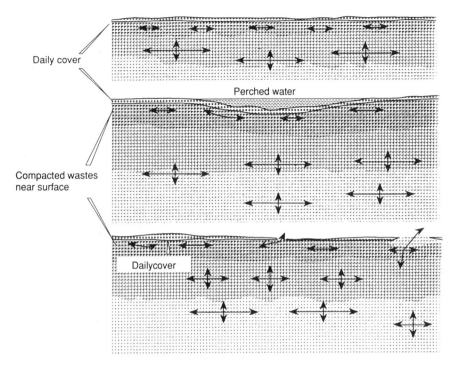

Figure 11.10 Pathways for gas and leachate movement within a landfill in which daily cover provides lateral permeability barriers to flow (adapted Department of the Environment, 1989).

gravel) or through joints and other fissures (hard-rock quarries), which may be accentuated by blasting. These sites may require the installation of a liner, depending in detail on the requirements of the operator and licensing authority.

Where a liner needs to be installed, a number of natural materials need to be considered (Harries-Rees, 1993):

- natural clay – if a site already has access to clay, this can be used to prepare a liner. It might occur as the natural base of the site (in which case care must be taken to ensure that there is an adequate vertical distance between the topographic base of the site and the base of the clay), or it might occur as a seam or overburden which can be stockpiled during mining. Natural clay is rarely moved into a site, unless a local source is available (e.g. the possibility of using waste material from other activities such as road construction). It is general to compact the clay base of a site, or to install a layer of clay 1 m thick.
- bentonite – many of the properties of bentonite which make it valuable as an industrial raw material are also those which make it valuable as a

lining material for a waste disposal operation. Sodium bentonite is almost exclusively used, either as Wyoming bentonite (predominantly in North America) or as sodium-exchanged (engineered) bentonite in Europe. Bentonite is not a cheap raw material, and so it is not laid down as a mineralogically-pure layer tens of centimetres thick. Instead, bentonite-enriched soil is prepared by mixing up to 8% bentonite with locally derived soil and is then laid down.
- composite bentonite–geotextile sandwiches – these represent a very effective way of using bentonite, by sandwiching it between two geotextiles to give a 'carpet' about 1 cm thick. It reduces the need to take up valuable airspace by installation of thick sealing layers, and can be produced under carefully controlled factory conditions.

Appendix A
Reference phase diagrams

APPENDIX A: REFERENCE PHASE DIAGRAMS

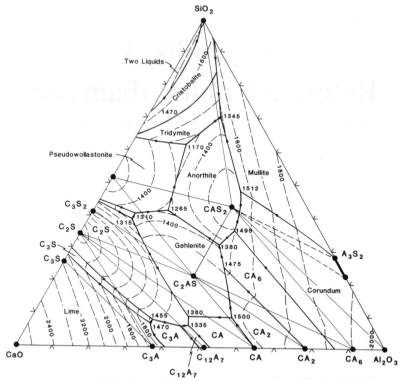

Figure A.1 System SiO_2–CaO–Al_2O_3 at one atmosphere (Levin, Robbins and McMurdie, 1964). This is the 'classic' cement system, and is also available for computer calculation using MTDATA. Temperatures of invariant points (i.e. where use of the phase rule shows that $f = 0$) are given in °C. The system includes the following crystalline phases:

Name	Formula	Abbreviated formula
Cristobalite	SiO_2	S
Tridymite	SiO_2	S
Pseudowollastonite	$CaSiO_3$	CS
Rankinite	$Ca_3Si_2O_7$	C_3S_2
Dicalcium silicate	Ca_2SiO_4	C_2S
Tricalcium silicate	Ca_3SiO_5	C_3S
Lime	CaO	C
Tricalcium aluminate	$Ca_3Al_2O_6$	C_3A
No shorthand name	$Ca_{12}Al_1O_{33}$	$C_{12}A_7$
Calcium aluminate	$CaAl_2O_4$	CA
Calcium dialuminate	$CaAl_4O_7$	CA_2
No shorthand name	$CaAl_{12}O_{19}$	CA_6
Corundum	Al_2O_3	A
Mullite[1]	$Al_6Si_2O_{13}$	A_3S_2
Anorthite	$CaAl_2Si_2O_8$	CAS_2
Gehlenite	$Ca_2Al_2SiO_7$	C_2AS

[1] shows solid solution towards both silica and alumina.

APPENDIX A: REFERENCE PHASE DIAGRAMS

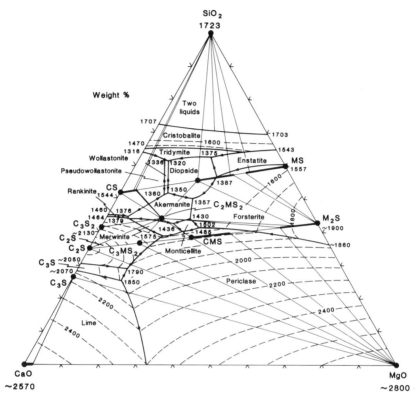

Figure A.2 System SiO_2–CaO–MgO at one atmosphere (Levin, Robbins and McMurdie, 1964). Temperatures of invariant points (i.e. where use of the phase rules shows that $f = 0$) are given in °C. The system includes the following crystalline phases:

Name	Formula	Abbreviated formula
Cristobalite	SiO_2	S
Tridymite	SiO_2	S
Pseudowollastonite[1]	$CaSiO_3$	CS
Rankinite	$Ca_3Si_2O_7$	C_3S_2
Dicalcium silicate	Ca_2SiO_4	C_2S
Tricalcium silicate	Ca_3SiO_5	C_3S
Lime	CaO	C
Magnesia (periclase)	MgO	M
Forsterite[2]	Mg_2SiO_4	M_2S
Enstatite[3]	$MgSiO_3$	MS
Diopside[4]	$CaMgSi_2O_6$	CMS_2
Akermanite	$Ca_2MgSi_2O_7$	C_2MS_2
Merwinite	$Ca_3MgSi_2O_8$	C_3MS_2
Monticellite[5]	$CaMgSiO_4$	CMS

[1] shows solid solution towards diopside; [2] shows solid solution towards monticellite; [3] shows solid solution towards diopside; [4] shows solid solution towards enstatite; [5] shows solid solution towards forsterite.

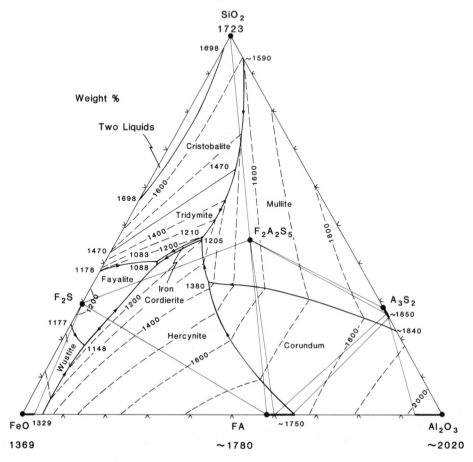

Figure A.3 System SiO_2–FeO–Al_2O_3 at one atmosphere (Levin, Robbins and McMurdie, 1964). Temperatures of invariant points (i.e. where use of the phase rule shows that $f = 0$) are given in °C. The systems includes the following crystalline phases:

Name	Formula	Abbreviated formula*
Cristobalite	SiO_2	S
Tridymite	SiO_2	S
Fayalite	Fe_2SiO_4	F_2S
Wüstite[1]	FeO	F
Hercynite[2]	$FeAl_2O_4$	FA
Corundum[3]	Al_2O_3	A
Mullite[4]	$Al_6Si_2O_{13}$	A_3S_2
Iron cordierite	$Fe_2Al_4Si_5O_{18}$	$F_2A_2S_5$

[1] shows solid solution towards hercynite; [2] shows solid solution towards alumina; [3] shows solid solution towards towards hercynite; [4] shows solid solution towards both silica and alumina; * note that all iron is as Fe^{2+}. In this case, F denotes FeO.

APPENDIX A: REFERENCE PHASE DIAGRAMS

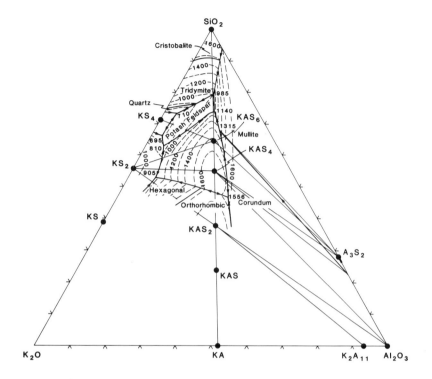

Figure A.4 System SiO_2–K_2O–Al_2O_3 at one atmosphere (Levin, Robbins and McMurdie, 1964). Note that large areas of this figure are left blank, reflecting the practical difficulties involved in working with potassium and aluminium-rich compositions, as well as the limited relevance of considering compositions consisting predominantly of K_2O, which do not occur in nature. Temperatures of invariant points (i.e. where use of the phase rule shows that $f = 0$) are given in °C. The system includes the following crystalline phases:

Name	Formula	Abbreviated formula
Cristobalite	SiO_2	S
Tridymite	SiO_2	S
Quartz	SiO_2	S
No shorthand name	$K_2Si_4O_9$	KS_4
Potassium disilicate	$K_2Si_2O_5$	KS_2
Potassium silicate	K_2SiO_3	KS
Potash	K_2O	K
Potassium aluminate	$K_2Al_2O_4$	KA
No shorthand name	$K_2Al_{22}O_{34}$	K_2A_{11}
Corundum	Al_2O_3	A
Mullite[1]	$Al_6Si_2O_{13}$	A_3S_2
Potassium aluminium silicate	$K_2Al_2SiO_6$	KAS
Kalsilite	$KAlSiO_4$	KAS_2
Leucite	$KAlSi_2O_6$	KAS_4
Potassium feldspar	$KAlSi_3O_8$	KAS_6

[1] shows solid solution towards both silica and alumina.

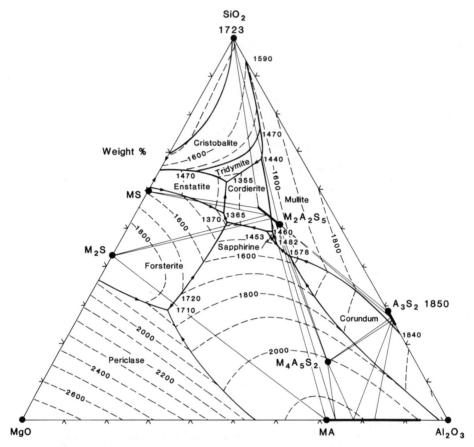

Figure A.5 System SiO_2–MgO–Al_2O_3 at one atmosphere (Levin, Robbins and McMurdie, 1964). Temperatures of invariant points (i.e. where use of the phase rules shows that $f = 0$) are given in °C. The system includes the following crystalline phases:

Name	Formula	Abbreviated formula
Cristobalite	SiO_2	S
Tridymite	SiO_2	S
Enstatite	$MgSiO_3$	MS
Forsterite	Mg_2SiO_4	M_2S
Magnesia	MgO	M
Spinel[1]	$MgAl_2O_4$	MA
Corundum	Al_2O_3	A
Mullite[2]	$Al_6Si_2O_{13}$	A_3S_2
Magnesium cordierite[3]	$Mg_2Al_4Si_5O_{18}$	$M_2A_2S_5$
Sapphirine	$Mg_4Al_{10}Si_2O_{23}$	$M_4A_5S_2$

[1] shows solid solution towards alumina; [2] shows solid solution towards both silica and alumina; [3] shows solid solution towards the MgO–SiO_2 join.

Appendix B
Detailed quality variation for the model evaluation used in Chapter 10

For the constrained deposit model summarized in Figure 10.15, the following histograms and tables respectively show variation in the following quality parameters: (a) thickness, (b) overburden, (c) brightness 1, (d) brightness 2, (e) flow, (f) viscosity, (g) titanium dioxide content, and (h) grit content (Figures B.1–B.8, Tables B.1–B.8).

APPENDIX B

Reserve Analysis

Thick

Z-Range		Area Acres/ Hectares	Reserves in Tonnes	Pct of Total- Volume	Cumulative	
From	To				Tonnes	Thick
3.00	6.00	2.4	259,861	1.7	259,861	5.49
6.00	9.00	8.4	1,285,914	8.6	1,545,775	7.33
9.00	12.00	13.9	2,903,606	19.3	4,449,380	9.17
12.00	15.00	13.6	3,590,109	23.9	8,039,489	10.66
15.00	18.00	8.8	2,894,525	19.3	10,934,014	11.78
18.00	21.00	6.2	2,360,318	15.7	13,294,332	12.66
21.00	24.00	3.4	1,520,779	10.1	14,815,111	13.25
24.00	27.00	0.4	220,700	1.5	15,035,811	13.34
27.00	30.00	0.0	0	0.0	15,035,811	13.34
30.00	33.00	0.0	0	0.0	15,035,811	13.34
3.00	33.00	57.2	15,035,811	100.0		

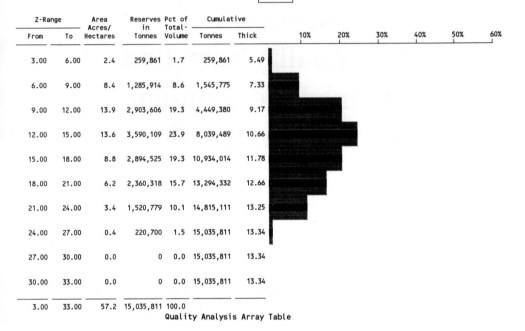

Quality Analysis Array Table

Thick

Z-Range		Areas in Hectares	Reserves in Tonnes	Pct of total- Volume	Weighted Average of Qualities for Each Band Split						
From	To				Thick	Burden	Brit1	Brit2	Flow	Viscos	Ti-O2	Grit%
3.0	6.0	2.4	259,861	1.7%	5.49	26.34	82.82	90.98	73.72	404.00	0.45	4.47
6.0	9.0	8.4	1,285,914	8.6%	7.87	20.48	82.47	90.31	74.26	333.59	0.59	3.75
9.0	12.0	13.9	2,903,606	19.3%	10.60	20.79	82.78	90.32	73.63	356.26	0.51	3.60
12.0	15.0	13.6	3,590,109	23.9%	13.35	22.37	83.30	90.40	73.64	389.07	0.46	3.41
15.0	18.0	8.8	2,894,525	19.3%	16.66	29.06	83.44	90.46	73.85	392.00	0.51	3.19
18.0	21.0	6.2	2,360,318	15.7%	19.34	26.97	83.63	90.60	74.16	402.09	0.54	3.01
21.0	24.0	3.4	1,520,779	10.1%	22.46	31.49	83.72	90.83	73.75	416.63	0.48	3.21
24.0	27.0	0.4	220,700	1.5%	25.50	27.49	83.87	90.78	74.09	427.10	0.38	1.57
27.0	30.0	0.0	0	0.0%	0.00	27.49	83.87	90.78	74.09	427.10	0.38	1.57
30.0	33.0	0.0	0	0.0%	0.00	27.49	83.87	90.78	74.09	427.10	0.38	1.57
3.0	33.0	57.2	15,035,811	100.0%								

DETAILED QUALITY VARIATION FOR MODEL EVALUATION

253

Reserve Analysis

Burden

Z-Range		Area Acres/ Hectares	Reserves in Tonnes	Pct of Total- Volume	Cumulative		10%	20%	30%	40%	50%	60%
From	To				Tonnes	Burden						
0.00	10.00	2.2	471,418	3.1	471,418	8.36						
10.00	20.00	17.5	4,237,843	28.2	4,709,260	14.69						
20.00	30.00	22.5	5,406,470	36.0	10,115,730	20.20						
30.00	40.00	15.1	4,920,085	32.7	15,035,814	24.02						
40.00	50.00	0.0	0	0.0	15,035,814	24.02						
50.00	60.00	0.0	0	0.0	15,035,814	24.02						
60.00	70.00	0.0	0	0.0	15,035,814	24.02						
70.00	80.00	0.0	0	0.0	15,035,814	24.02						
80.00	90.00	0.0	0	0.0	15,035,814	24.02						
90.00	100.00	0.0	0	0.0	15,035,814	24.02						
0.00	100.00	57.2	15,035,814	100.0								

Quality Analysis Array Table

Burden

Z-Range		Areas in Hectares	Reserves in Tonnes	Pct of total- Volume	Weighted Average of Qualities for Each Band Split							
From	To				Thick	Burden	Brit1	Brit2	Flow	Viscos	Ti-O2	Grit%
0.0	10.0	2.2	471,418	3.1%	11.46	8.36	82.14	90.08	73.39	400.22	0.60	4.98
10.0	20.0	17.5	4,237,843	28.2%	13.43	15.48	82.92	90.12	73.78	376.69	0.52	3.01
20.0	30.0	22.5	5,406,470	36.0%	13.75	25.01	83.33	90.61	74.08	361.06	0.50	3.05
30.0	40.0	15.1	4,920,085	32.7%	17.73	34.71	83.55	90.69	73.64	414.56	0.50	3.80
40.0	50.0	0.0	0	0.0%	0.00	34.71	83.55	90.69	73.64	414.56	0.50	3.80
50.0	60.0	0.0	0	0.0%	0.00	34.71	83.55	90.69	73.64	414.56	0.50	3.80
60.0	70.0	0.0	0	0.0%	0.00	34.71	83.55	90.69	73.64	414.56	0.50	3.80
70.0	80.0	0.0	0	0.0%	0.00	34.71	83.55	90.69	73.64	414.56	0.50	3.80
80.0	90.0	0.0	0	0.0%	0.00	34.71	83.55	90.69	73.64	414.56	0.50	3.80
90.0	100.0	0.0	0	0.0%	0.00	34.71	83.55	90.69	73.64	414.56	0.50	3.80
0.0	100.0	57.2	15,035,814	100.0%								

APPENDIX B

Reserve Analysis

Brit1

Z-Range		Area Acres/ Hectares	Reserves in Tonnes	Pct of Total-Volume	Cumulative		10%	20%	30%	40%	50%	60%
From	To				Tonnes	Brit1						
79.20	80.00	0.4	77,913	0.5	77,913	79.64						
80.00	80.80	0.8	124,174	0.8	202,087	80.16						
80.80	81.60	2.5	405,712	2.7	607,799	80.91						
81.60	82.40	7.8	1,598,167	10.6	2,205,966	81.77						
82.40	83.20	16.7	4,329,051	28.8	6,535,017	82.48						
83.20	84.00	20.3	5,734,602	38.1	12,269,618	83.00						
84.00	84.80	8.0	2,503,743	16.7	14,773,361	83.22						
84.80	85.60	0.8	262,452	1.7	15,035,813	83.25						
85.60	86.40	0.0	0	0.0	15,035,813	83.25						
79.20	86.40	57.2	15,035,813	100.0								

Quality Analysis Array Table

Brit1

Z-Range		Areas in Hectares	Reserves in Tonnes	Pct of total-Volume	Weighted Average of Qualities for Each Band Split							
From	To				Thick	Burden	Brit1	Brit2	Flow	Viscos	Ti-O2	Grit%
79.2	80.0	0.4	77,913	0.5%	10.52	12.72	79.64	90.44	73.90	323.81	0.56	3.47
80.0	80.8	0.8	124,174	0.8%	8.63	14.94	80.49	90.33	74.50	338.63	0.58	2.52
80.8	81.6	2.5	405,712	2.7%	8.64	18.36	81.28	90.25	74.44	349.95	0.58	2.63
81.6	82.4	7.8	1,598,167	10.6%	11.26	17.52	82.10	90.07	74.00	332.03	0.60	3.85
82.4	83.2	16.7	4,329,051	28.8%	14.35	24.45	82.84	90.29	73.74	378.08	0.52	3.52
83.2	84.0	20.3	5,734,602	38.1%	15.94	27.50	83.60	90.66	73.75	401.25	0.47	3.16
84.0	84.8	8.0	2,503,743	16.7%	16.78	26.37	84.27	90.68	73.81	399.63	0.48	3.36
84.8	85.6	0.8	262,452	1.7%	19.05	29.38	85.03	90.80	74.89	375.46	0.55	2.61
85.6	86.4	0.0	0	0.0%	0.00	29.38	85.03	90.80	74.89	375.46	0.55	2.61
79.2	86.4	57.2	15,035,813	100.0%								

DETAILED QUALITY VARIATION FOR MODEL EVALUATION

Reserve Analysis

Brit2

Z-Range		Area Acres/ Hectares	Reserves in Tonnes	Pct of Total-Volume	Cumulative		10%	20%	30%	40%	50%	60%
From	To				Tonnes	Brit2						
88.00	88.40	0.0	0	0.0	0	0.00						
88.40	88.80	0.3	37,534	0.2	37,534	88.74						
88.80	89.20	1.0	180,116	1.2	217,650	88.97						
89.20	89.60	2.9	691,193	4.6	908,843	89.34						
89.60	90.00	8.3	2,146,103	14.3	3,054,946	89.66						
90.00	90.40	12.3	3,128,183	20.8	6,183,129	89.94						
90.40	90.80	16.5	4,465,319	29.7	10,648,448	90.22						
90.80	91.20	11.2	3,213,620	21.4	13,862,068	90.40						
91.20	91.60	3.8	877,024	5.8	14,739,092	90.46						
91.60	92.00	1.0	296,721	2.0	15,035,813	90.48						
88.00	92.00	57.2	15,035,813	100.0								

Quality Analysis Array Table

Brit2

Z-Range		Areas in Hectares	Reserves in Tonnes	Pct of total-Volume	Weighted Average of Qualities for Each Band Split							
From	To				Thick	Burden	Brit1	Brit2	Flow	Viscos	Ti-O2	Grit%
88.0	88.4	0.0	0	0.0%	0.00	29.38	85.03	90.80	74.89	375.46	0.55	2.61
88.4	88.8	0.3	37,534	0.2%	7.65	18.60	82.50	88.74	75.08	368.72	0.80	3.47
88.8	89.2	1.0	180,116	1.2%	9.28	15.99	82.69	89.02	74.52	398.72	0.62	3.72
89.2	89.6	2.9	691,193	4.6%	12.68	18.01	82.93	89.46	73.59	393.74	0.53	3.32
89.6	90.0	8.3	2,146,103	14.3%	14.29	19.52	82.98	89.80	73.74	390.61	0.56	3.19
90.0	90.4	12.3	3,128,183	20.8%	14.29	23.78	82.89	90.21	73.84	370.11	0.56	3.76
90.4	90.8	16.5	4,465,319	29.7%	15.12	26.40	83.32	90.60	73.77	371.32	0.50	3.40
90.8	91.2	11.2	3,213,620	21.4%	16.34	28.10	83.64	90.99	73.99	403.37	0.45	3.05
91.2	91.6	3.8	877,024	5.8%	13.82	28.86	83.68	91.38	73.89	394.79	0.43	2.87
91.6	92.0	1.0	296,721	2.0%	19.19	32.93	83.63	91.70	73.56	412.10	0.41	3.68
88.0	92.0	57.2	15,035,813	100.0%								

APPENDIX B

Reserve Analysis

Flow

Z-Range		Area Acres/ Hectares	Reserves in Tonnes	Pct of Total-Volume	Cumulative		10%	20%	30%	40%	50%	60%
From	To				Tonnes	Flow						
70.20	70.80	0.2	45,412	0.3	45,412	70.72						
70.80	71.40	0.7	175,488	1.2	220,900	71.04						
71.40	72.00	3.3	834,505	5.6	1,055,405	71.62						
72.00	72.60	5.8	1,495,915	9.9	2,551,320	72.02						
72.60	73.20	5.7	1,389,525	9.2	3,940,845	72.33						
73.20	73.80	5.6	1,542,968	10.3	5,483,813	72.67						
73.80	74.40	15.4	4,574,786	30.4	10,058,598	73.34						
74.40	75.00	15.7	3,758,345	25.0	13,816,943	73.71						
75.00	75.60	4.5	1,124,231	7.5	14,941,174	73.82						
75.60	76.20	0.3	94,637	0.6	15,035,811	73.83						
76.20	76.80	0.0	0	0.0	15,035,811	73.83						
70.20	76.20	57.2	15,035,811	100.0								

Quality Analysis Array Table

Flow

Z-Range		Areas in Hectares	Reserves in Tonnes	Pct of total-Volume	Weighted Average of Qualities for Each Band Split						
From	To				Thick	Burden	Brit1	Brit2	Flow	Viscos	Ti-O2	Grit%
70.2	70.8	0.2	45,412	0.3%	13.80	14.07	84.37	90.38	70.72	547.48	0.24	7.57
70.8	71.4	0.7	175,488	1.2%	12.41	19.04	83.41	90.10	71.13	354.01	0.40	4.54
71.4	72.0	3.3	834,505	5.6%	15.57	28.54	83.22	90.41	71.77	388.92	0.42	3.80
72.0	72.6	5.8	1,495,915	9.9%	14.17	27.56	83.46	90.66	72.31	400.88	0.43	4.14
72.6	73.2	5.7	1,389,525	9.2%	13.17	20.07	83.06	90.33	72.89	387.06	0.47	3.81
73.2	73.8	5.6	1,542,968	10.3%	15.18	25.27	83.29	90.55	73.53	363.94	0.53	3.10
73.8	74.4	15.4	4,574,786	30.4%	16.29	26.85	83.24	90.48	74.15	404.79	0.52	3.27
74.4	75.0	15.7	3,758,345	25.0%	13.83	23.97	83.09	90.43	74.68	372.03	0.53	3.03
75.0	75.6	4.5	1,124,231	7.5%	15.04	22.20	83.62	90.57	75.22	342.46	0.55	2.79
75.6	76.2	0.3	94,637	0.6%	18.98	18.92	84.09	91.06	75.73	329.05	0.58	1.87
76.2	76.8	0.0	0	0.0%	0.00	18.92	84.09	91.06	75.73	329.05	0.58	1.87
70.2	76.2	57.2	15,035,811	100.0%								

DETAILED QUALITY VARIATION FOR MODEL EVALUATION 257

Reserve Analysis

Viscos

Z-Range		Area Acres/ Hectares	Reserves in Tonnes	Pct of Total-Volume	Cumulative		10%	20%	30%	40%	50%	60%
From	To				Tonnes	Viscos						
160.00	200.00	0.1	14,851	0.1	14,851	180.04						
200.00	240.00	1.5	356,157	2.4	371,008	222.10						
240.00	280.00	3.5	715,785	4.8	1,086,793	248.64						
280.00	320.00	7.6	1,650,763	11.0	2,737,556	280.53						
320.00	360.00	10.3	2,385,301	15.9	5,122,857	308.76						
360.00	400.00	11.3	3,091,062	20.6	8,213,919	336.08						
400.00	440.00	11.5	3,747,078	24.9	11,960,997	362.57						
440.00	480.00	8.6	2,286,658	15.2	14,247,655	377.80						
480.00	520.00	2.3	651,496	4.3	14,899,151	382.82						
520.00	560.00	0.5	119,336	0.8	15,018,487	383.99						
560.00	600.00	0.1	17,325	0.1	15,035,812	384.20						
160.00	600.00	57.2	15,035,812	100.0								

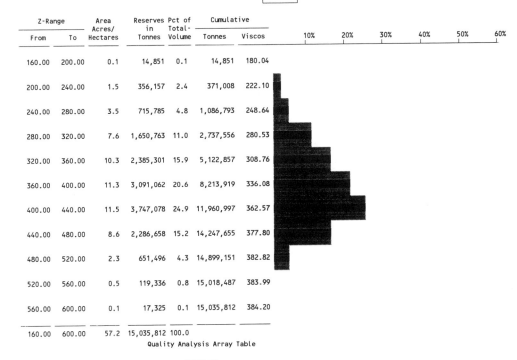

Quality Analysis Array Table

Viscos

Z-Range		Areas in Hectares	Reserves in Tonnes	Pct of total-Volume	Weighted Average of Qualities for Each Band Split							
From	To				Thick	Burden	Brit1	Brit2	Flow	Viscos	Ti-O2	Grit%
160.0	200.0	0.1	14,851	0.1%	12.00	20.00	82.05	90.40	74.65	180.04	0.44	3.10
200.0	240.0	1.5	356,157	2.4%	13.27	21.26	82.44	90.40	73.41	223.85	0.56	4.21
240.0	280.0	3.5	715,785	4.8%	11.64	22.44	83.03	90.25	73.68	262.39	0.62	3.22
280.0	320.0	7.6	1,650,763	11.0%	12.17	20.51	82.95	90.31	74.09	301.53	0.58	2.96
320.0	360.0	10.3	2,385,301	15.9%	12.68	21.30	82.87	90.41	74.12	341.16	0.53	2.92
360.0	400.0	11.3	3,091,062	20.6%	15.30	26.58	83.24	90.57	73.93	381.35	0.49	3.32
400.0	440.0	11.5	3,747,078	24.9%	17.69	29.10	83.58	90.71	73.72	420.62	0.47	3.29
440.0	480.0	8.6	2,286,658	15.2%	15.17	25.36	83.41	90.30	73.74	457.47	0.49	3.81
480.0	520.0	2.3	651,496	4.3%	15.66	23.72	83.58	90.41	73.41	492.68	0.45	3.90
520.0	560.0	0.5	119,336	0.8%	13.86	17.72	83.77	90.32	72.05	530.13	0.29	4.96
560.0	600.0	0.1	17,325	0.1%	14.00	14.00	84.50	90.45	70.68	564.61	0.22	7.92
160.0	600.0	57.2	15,035,812	100.0%								

Reserve Analysis

Ti-02

Z-Range		Area Acres/ Hectares	Reserves in Tonnes	Pct of Total-Volume	Cumulative		10%	20%	30%	40%	50%	60%
From	To				Tonnes	Ti-02						
0.10	0.30	2.1	335,486	2.2	335,486	0.24						
0.30	0.50	25.2	7,322,010	48.7	7,657,495	0.42						
0.50	0.70	26.8	6,650,970	44.2	14,308,464	0.49						
0.70	0.90	2.6	655,520	4.4	14,963,984	0.50						
0.90	1.10	0.3	38,262	0.3	15,002,246	0.50						
1.10	1.30	0.3	33,569	0.2	15,035,815	0.51						
1.30	1.50	0.0	0	0.0	15,035,815	0.51						
0.10	1.40	57.2	15,035,815	100.0								

Quality Analysis Array Table

Ti-02

Z-Range		Areas in Hectares	Reserves in Tonnes	Pct of total-Volume	Weighted Average of Qualities for Each Band Split							
From	To				Thick	Burden	Brit1	Brit2	Flow	Viscos	Ti-02	Grit%
0.1	0.3	2.1	335,486	2.2%	9.75	20.48	83.83	90.72	71.79	492.92	0.24	4.98
0.3	0.5	25.2	7,322,010	48.7%	15.90	26.84	83.47	90.69	73.64	399.92	0.43	3.19
0.5	0.7	26.8	6,650,970	44.2%	14.03	23.24	83.02	90.29	74.12	364.34	0.58	3.33
0.7	0.9	2.6	655,520	4.4%	15.83	24.10	82.98	90.09	73.99	367.56	0.74	4.51
0.9	1.1	0.3	38,262	0.3%	7.79	24.63	82.54	89.22	74.74	272.91	0.99	2.31
1.1	1.3	0.3	33,569	0.2%	6.80	24.62	82.28	88.78	75.37	255.08	1.15	1.84
1.3	1.5	0.0	0	0.0%	0.00	24.62	82.28	88.78	75.37	255.08	1.15	1.84
0.1	1.4	57.2	15,035,815	100.0%								

DETAILED QUALITY VARIATION FOR MODEL EVALUATION

259

Reserve Analysis

Grit%

Z-Range		Area Acres/ Hectares	Reserves in Tonnes	Pct of Total-Volume	Cumulative		10%	20%	30%	40%	50%	60%
From	To				Tonnes	Grit%						
0.00	2.00	8.7	2,437,459	16.2	2,437,459	1.47						
2.00	4.00	30.1	8,139,482	54.1	10,576,941	2.69						
4.00	6.00	15.8	4,014,814	26.7	14,591,755	3.23						
6.00	8.00	2.2	401,655	2.7	14,993,410	3.33						
8.00	10.00	0.3	36,907	0.2	15,030,317	3.34						
10.00	12.00	0.0	5,496	0.0	15,035,813	3.34						
0.00	12.00	57.2	15,035,813	100.0								

Quality Analysis Array Table

Grit%

Z-Range		Areas in Hectares	Reserves in Tonnes	Pct of total-Volume	Weighted Average of Qualities for Each Band Split							
From	To				Thick	Burden	Brit1	Brit2	Flow	Viscos	Ti-O2	Grit%
0.0	2.0	8.7	2,437,459	16.2%	16.19	22.14	83.47	90.47	74.07	358.70	0.51	1.47
2.0	4.0	30.1	8,139,482	54.1%	14.98	24.82	83.21	90.54	74.00	385.20	0.50	3.06
4.0	6.0	15.8	4,014,814	26.7%	14.53	27.47	83.21	90.36	73.47	397.55	0.52	4.67
6.0	8.0	2.2	401,655	2.7%	9.71	20.39	83.29	90.49	72.53	401.28	0.44	6.69
8.0	10.0	0.3	36,907	0.2%	6.73	26.07	82.26	89.90	75.18	236.30	0.71	8.98
10.0	12.0	0.0	5,496	0.0%	6.00	28.00	82.10	89.70	75.20	200.03	0.75	11.20
0.0	12.0	57.2	15,035,813	100.0%								

Appendix C
Computer hardware and software

A.1 RESERVE EVALUATION SOFTWARE

We selected the PC/Cores reserve evaluation and mine planning software system as the basis for this chapter for several reasons: (1) PC/Cores is currently in use at the University of Manchester as a teaching tool; (2) PC/Cores is a remarkably easy software package to learn and to use; (3) it is currently in use in many countries around the world and is being used in the mining of a wide range of minerals, including various clays, coal, limestone, marble, and others. [For details of the availability of a demonstration disk based on the material used in Chapter 10, contact Mentor Consultants or the author.]

PC/Cores is a trademark belonging to Mentor Consultants Inc., of the United States, who are the designers, developers, and owners of the system. Further information can be obtained from Mentor Consultants, Inc., Wheaton, Illinois, USA.

A.2 COMPUTER HARDWARE

PC/Cores operates on any standard IBM-compatible personal computer, that is, any PC which is based on the Intel microchips (80286, 80386, 80486, Pentium, etc.). In general, it will not operate on Apple Macintosh computers (unless on a powerful machine with PC emulation), nor on the so-called 'work stations' or Sun-type machines. It operates under either the DOS or Windows operating systems.

A typical lower-cost computer configuration would include the following components and attachments: (1) An 80386/80486 computer with a hard disk drive and at least one floppy drive, along with a VGA monitor and at least 4 megabytes of RAM memory; (2) a dot matrix or laser printer; (3) a digitizing tablet, even if only a small one; (4) a pen plotter for drawing maps and borehole profiles.

References

Adams, J.M. (1993) Particle size and shape effects in materials science: examples from polymer and paper systems. *Clay Minerals*, **28**, 509–30.

Alderton, D.H.M. (1993) Granite-associated mineralisation in south-west England, in *Mineralization in the British Isles* (eds R.A.D. Pattrick and D.A. Polya), Chapman & Hall, London, 270–354.

Annels, A.E. (ed.) (1992) *Case histories and methods in mineral resource evaluation*. Geological Society Special Publication 63, Geological Society of London, 313pp.

Ansems, R. (1990) Economic evaluation of glass raw materials. *Industrial Minerals*, **273** (June 1990), 57–63.

Archer, A.A. (1972) *Sand and Gravel as Aggregate, Mineral Resources Consultative Committee, Mineral Dossier No. 4*, Her Majesty's Stationery Office, London, 29pp.

Austin, G.T. (1984) *Shreve's Chemical Process Industries* (5th ed). McGraw-Hill International Editions, New York, 859pp.

Barnes, P. (ed) (1983) *Structure and performance of cements*, Appled Science Publishers, Barking.

Blakey, N.C. and Maris, P.J. (1990) Methane recovery from the anaerobic digestion of landfill leachate. *Department of Energy Contractor Report ETSU B 1223*, Her Majesty's Stationery Office, 48pp.

Bogue, R.H. (1955) *The Chemistry of Portland Cement*, Reinhold, New York, 793pp.

Bowen, N.L. (1956) *The Evolution of the Igneous Rocks*, Dover Publications Inc., New York, 334pp.

Bradley, W.H. (1948) Geology of Green River formation and associated Eocene rocks in southwestern Wyoming and adjacent parts of Colorado and Utah. *United States Geological Survey Professional Paper*, **496-A**, 86pp.

Bray, C.J. and Spooner, E.T.C. (1983) Sheeted vein Sn-W mineralization and greisenization associated with economic kaolinization, Goonbarrow china clay pit, St. Austell, Cornwall, England: geologic relationships and geochronology. *Economic Geology*, **78**, 1064–89.

Bristow, C.M. (1977) A review of the evidence for the origin of the kaolin deposits in S.W. England, in E. Galan (ed.) Proceedings of the 8th International Kaolin Symposium and Meeting on Alunite, Madrid–Rome, September 7–16, 1977, No. K-2, 19pp.

REFERENCES

Bristow, C.M. (1987) Society's changing requirements for primary raw materials. *Industrial Minerals*, **232** (January 1987), 59–65.

Bristow, C.M. (1989) World kaolins. Genesis, exploitation, and application, in *Industrial clays, a special review* (ed. G. Clarke) Metal Bulletin plc, pp. 8–17.

Bristow, C.M. and Exley, C.S. (1994) Historical and geological aspects of the china clay industry of south-west England. *Transactions of the Royal Geological Society of Cornwall*, **21**, 247–314.

Brown, T.H., Berman, R.G. and Perkins, E.H. (1988) GEOCALC II: PTA-SYSTEM: *Software for the Calculation and Display of Pressure-Temperature-Activity Phase Diagrams*, University of British Columbia, Department of Geological Sciences, Vancouver, British Columbia.

Burt, R.O., Flemming, J., Simard, R. and Vanstone, P.J. (1988) Tanco – a new name in low iron spodumene. *Industrial Minerals*, **244** (January 1988), 53–9.

Bye, G.C. (1983) *Portland Cement. Composition, Production and Properties*, Pergamon, Oxford, 149pp.

Chesters, J.H. (1983) *Refractories. Production and Properties*, Metals Society, London, 553pp.

Christidis, G. and Scott, P.W. (1993) Laboratory evaluation of bentonites. *Industrial Minerals*, **311** (August 1993), 51–7.

Clark, A.M. (1993) *Hey's Mineral Index*, Chapman & Hall, London, 782pp.

Conde-Pumpido, R., Ferrón, J.J. and Campillo, G. (1988) Influence of granulometric factors on the rheology of kaolins of Galicia, Spain. *Applied Clay Science*, **3**, 177–85.

Cook, P.J. (1984) Spatial and temporal controls on the formation of phosphate deposits – a review, in *Phosphate Minerals* (eds J.O. Nriagu and P.B. Moore), Springer Verlag, Berlin, 242–74.

Coultate, T.P. (1992) *The Chemistry of Food and its Components*, Royal Society of Chemistry, London, 325pp.

Crouch, F. (1993) Shipping practices. *Industrial Minerals*, **312** (September 1993), 39–47.

Crouse, R.A., Cerny, P., Trueman, D.L. and Burt, R.O. (1979) The Tanco pegmatite, southeastern Manitoba. *Canadian Mining and Metallurgy Bulletin*, February 1979, 142–51.

Daniel, D. (ed.) (1993a) *Geotechnical Practice for Waste Disposal*, Chapman & Hall, London, 683pp.

Daniel, D. (1993b) Landfills and impoundments, in *Geotechnical Practice for Waste Disposal* (ed. D. Daniel), Chapman & Hall, London, 97–112.

Deardorff, D.L. and Mannion, L.E. (1971) Wyoming trona deposits. *Contributions of Geology* (ed. R.B. Parker), **10/1**, University of Wyoming, Laramie, 25–39.

Deer, W.A., Howie, R.A. and Zussman, J. (1992) *An Introduction to the Rock-forming Minerals*, Longman, London, 696pp.

Department of the Environment (1986) Landfilling wastes, *Waste Management Paper No. 26*, Her Majesty's Stationery Office, 206pp.

Department of the Environment (1989) Landfill gas. *Waste Management Paper No. 27*, Her Majesty's Stationery Office, 82pp.

Department of the Environment (1991a) Minerals Planning Guidance: Provision of raw material for the cement industry. *Minerals Planning Guidance*, Note 10, 82pp.

Department of the Environment (1991b) *Environmental Effects of Surface Mineral Workings*, Her Majesty's Stationery Office, 176pp.

Dickson, E.M. (1986) *Raw Materials for the Refractories Industry. Industrial Minerals Consumer Survey*, Metal Bulletin Journals Ltd, London, 198pp.

Doremus, R.H. (1973) *Glass science*, Wiley.

Dunham, A.C. (1992) Developments in industrial mineralogy: I. The mineralogy of brick making, *Proceedings of the Yorkshire Geological Society*, **49**, 95–104.

Earp, J.R. and Taylor, B.J. (1986) *Geology of the Country around Chester and Winsford. Memoir of the British Geological Survey, sheet 109*, Her Majesty's Stationery Office, London, 119pp.

Ehlers, E.G. (1972) *The Interpretation of Geological Phase Diagrams*, W.H. Freeman, San Francisco, 280pp.

Ehrenberg, S.N., Aagaard, P., Wilson, M.J. et al. (1993) Depth-dependent transformation of kaolinite to dickite in sandstones of the Norwegian continental shelf. *Clay Minerals*, **28**, 325–52.

Esteoule-Choux, J. (1983) Kaolinitic weathering profiles in Brittany: genesis and economic importance, in *Residual Deposits: Surface Related Weathering Processes and Methods* (ed. R.C.L. Wilson), Geological Society of London, Special Publication No. 11, 33–8.

Eugster, H.P. and Wones, D.R. (1962) Stability relations of the ferruginous biotite, annite. *Journal of Petrology*, **3**, 82–125.

Floyd, P.A., Exley, C.S. and Styles, M.T. (1992) *Igneous Rocks of South-west England*, Chapman & Hall, London, 272pp.

Garrels, R.M. (1984) Montmorillonite/illite stability diagrams. *Clays and Clay Minerals*, **32**, 161–6.

Garrett, D.E. (1992) *Natural soda ash. Occurrences, processing and use*, Van Nostrand Reinhold, New York, 636pp.

Griffiths, J. (1989) Olivine. *Industrial Minerals*, **256** (January 1989), 25–35.

Hamilton, D.L. and Henderson, C.M.B. (1968) The preparation of silicate compositions by a gelling method, *Mineralogical Magazine*, **36**, 832–8.

Hammersley, G.P. (1989) The use of petrography in the evaluation of aggregates. *Concrete*, **23**, 10.

Harben, P.W. (1977) *Raw Materials for the Glass Industry. Industrial Minerals Consumer Survey*, Metal Bulletin plc, London.

Harben, P.W. and Bates, R.L. (1990) *Industrial Minerals. Geology and World Deposits*, Metal Bulletin plc, London, 312pp.

Harries-Rees, K. (1992) Minerals in detergents. *Industrial Minerals*, **302** (November 1992), 37–49.

Harries-Rees, K. (1993) Minerals in waste and effluent treatment. *Industrial Minerals*, **308** (May 1993), 29–39.

Harris, P.M. (1977a) *Sandstone. Mineral Resources Consultative Committee, Mineral Dossier No. 17*, Her Majesty's Stationery Office, London, 37pp.

Harris, P.M. (1977b) *Igneous and Metamorphic Rock. Mineral Resources Consultative Committee, Mineral Dossier No. 19*, Her Majesty's Stationery Office, London, 62pp.

REFERENCES

Harris, P.M. (1982) *Limestone and Dolomite. Mineral Resources Consultative Committee, Mineral Dossier No. 23*, Her Majesty's Stationery Office, London, 111pp.

Harrison, D.J. and Adlam, K.A.McL. (1985) *Limestones of the Peak. Mineral Assessment Report 144*, British Geological Survey, Her Majesty's Stationery Office, London, 40pp.

Highley, D.E. (1975) *Ball Clay. Mineral Resources Consultative Committee, Mineral Dossier No. 11*, Her Majesty's Stationery Office, London, 32pp.

Highley, D.E. (1977) *Silica. Mineral Resources Consultative Committee, Mineral Dossier No. 18*, Her Majesty's Stationery Office, London, 57pp.

Highley, D.E. (1982) *Fireclay. Mineral Resources Consultative Committee, Mineral Dossier No. 24*, Her Majesty's Stationery Office, London, 72pp.

Highley, D.E. (1984) *China Clay. Mineral Resources Consultative Committee, Mineral Dossier No. 26*, Her Majesty's Stationery Office, London, 65pp.

Hower, J., Eslinger, E.V., Hower, M.E. and Perry, E.A. (1976) Mechanism of burial metamorphism of argillaceous sediment: 1. Mineralogical and chemical evidence. *Geological Society of America Bulletin*, **87**, 725–37.

Inan, K., Dunham, A.C. and Esson, J. (1973) Mineralogy, chemistry and origin of Kirka borate deposit, Eskishehir Province, Turkey. *Transactions of the Institution of Mining and Metallurgy, Section B*, **82**, B114–B123.

Intarapravich, D. (1992) Industrial minerals and the Thai economy. *Industrial Minerals*, **296** (May 1992), 107–19.

Isaaks, E.H. and Srivastava, R. (1989) *An Introduction to Applied Geostatistics*, Oxford University Press, Oxford, 561pp.

Jackson, N.J., Willis-Richards, J., Manning, D.A.C. and Sams, M.S. (1989) Evolution of the Cornubrian ore field, southwest England: Part II. Mineral deposits and ore-forming processes. *Economic Geology*, **84**, 1101–33

Jennings, B.R. (1993) Size and thickness measurement of polydisperse clay samples. *Clay Minerals*, **28**, 485–94.

Jepson, W.B. (1984) Kaolins: their properties and uses. *Philosophical Transactions of the Royal Society of London*, **A311**, 411–32.

Kimyongur, N. and Scott, P.W. (1986) Calcined natural magnesite – influence of time and temperature on the transformation and resulting industrial properties. *Materials Science Forum*, **7**, 83–90.

Kingsnorth, D.J. (1988) Lithium minerals in glass. New Directions. *Industrial Minerals*, **244** (January 1988), 49–52.

Lea, F.M. (1970) *The Chemistry of Cement and Concrete*, Edward Arnold, London, 727pp.

Lefond, S.J. (editor in chief) (1983) *Industrial Minerals and Rocks* (5th edn), Society of Mining Engineers, American Institute of Mining, Metallurgical, and Petroleum Engineers, Inc., New York, 2 vols, 1446pp.

Levin, E.M., McMurdie, H.F. and Hall, F.P. (1956) *Phase Diagrams for Ceramists*, American Ceramic Society, 2nd edition, Columbus, Ohio, 286pp.

Levin, E.M., Robbins, C.R. and McMurdie, H.F. (1964) *Phase Diagrams for Ceramists*. American Ceramic Society, Columbus, Ohio, 601pp.

Lofty, G.J., Hillier, J.A., Cooke, S.A., Linley, K.A. and Singh, H.R. (1992) *World Mineral Statistics 1986–1990. Volume 2: Industrial Minerals*, British Geological Survey, Her Majesty's Stationery Office, London, 137pp.

London, D. (1984) Experimental phase equilibria in the system $LiAlSiO_4$-SiO_2-H_2O: a petrogenetic grid for lithium-rich pegmatites. *American Mineralogist*, **69**, 995–1004.

Macaulay, C.I., Fallick, A.E. and Haszeldine, R.S. (1993) Textural and isotopic variations in diagenetic kaolinite from the Magnus Oilfield sandstones. *Clay Minerals*, **28**, 625–39.

Manning, D.A.C. and Exley, C.S. (1984) The origins of late-stage rocks in the St Austell granite – a re-interpretation. *Journal of the Geological Society*, **141**, 581–591.

Manning, D.A.C., Gestsdóttir, K. and Rae, E.I.C. (1992) Feldspar dissolution in the presence of organic acid anions under diagenetic conditions: an experimental study, *Organic Geochemistry*, **19**, 483–492.

McMichael, B. (1989) Chromite. *Industrial Minerals*, **257** (February 1989), 25–45.

McVey, H. (1994) The importance of industrial minerals in everyday life. *Geology Today*, **10**, 12–13.

Mew, M.C. (1980) *World Survey of Phosphate Deposits*. British Sulphur Corporation, London, 237pp.

Moorlock, B.S.P. and Highley, D.E. (1991) An appraisal of fuller's earth resources in England and Wales. *British Geological Survey Technical Report WA/92/75*, 87pp.

Muan, A. and Osborn, E.F. (1965) *Phase Equilibria among Oxides in Steel Making*, Addison Wesley, Reading MA, 236pp.

Notholt, A.J.G. (1979) The economic geology and development of igneous phosphate deposits in Europe and the USSR. *Economic Geology*, **74**, 339–50.

Notholt, A.J.G. and Highley, D.E. (1975) *Gypsum and Anhydrite. Mineral Resources Consultative Committee, Mineral Dossier No. 13*, Her Majesty's Stationery Office, London 38pp.

O'Driscoll, M. (1988) Dolomite. *Industrial Minerals*, **252** (September 1988), 37–63

O'Driscoll, M. (1989) Bentonite: overcapacity in need of markets, in *Industrial Clays* (ed. G. Clarke), Metal Bulletin plc, 55–71.

Odom, I.E. (1984) Smectite clay minerals: properties and uses. *Philosophical Transactions of the Royal Society of London*, **A311**, 391–409.

Pearson, W.J. (1962) Salt deposits of Canada. *Proceedings of the First Symposium on Salt*, Northern Ohio Geological Society, Cleveland, Ohio, 197–239.

Prentice, J.E. (1988) Evaluation of brick-clay reserves. *Transactions of the Institution of Mining and Metallurgy (Section B: Applied Earth Science)*, **97**, 9–14.

Prentice, J.E. (1990) *Geology of Construction Materials*, Chapman & Hall, 202pp.

Ridgeway, J.M. (1982) *Common Clay and Shale. Mineral Resources Consultative Committee, Mineral Dossier No. 22*, Her Majesty's Stationery Office, London, 164pp.

Rimstidt, J.D. and Barnes, H.L. (1980) The kinetics of silica-water reactions. *Geochimica et Cosmochimica Acta*, **44**, 1683–99.

Robertson, I.M.D. and Eggleton, R.A. (1991) Weathering of granitic muscovite to kaolinite and halloysite and of plagioclase-derived kaolinite to halloysite. *Clays and Clay Minerals*, **39**, 113–26.

Robinson, H.D. (1990) Leachate composition and treatment, in *The 1980s. A Decade of Progress? Achievements in Waste Management and Research*, Proceedings of the 1990 Harwell Waste Management Symposium, AEA Environment and Energy, Harwell, Oxon, 44–52.

Searle, A.F. (1912) *An Introduction to British Clays, Shales and Sands*, Griffin and Co., London.

Selwood, E.B., Edwards, R.A., Simpson, S., Chesher, J.A. et al. (1984) *Geology of the Country around Newton Abbot. Memoir of the British Geological Survey, Sheet 399.* Her Majesty's Stationery Office, London, 212pp.

Sheppard, S.M.F. (1977) The Cornubrian batholith, S.W. England: D/H and $^{18}O/^{16}O$ studies of kaolinite and other alteration minerals. *Journal of the Geological Society of London*, **133**, 573–91.

Slepetys, R.A. and Cleland, A.J. (1993) Determination of shape of kaolin pigment particles. *Clay Minerals*, **28**, 495–508.

Smith, M.R. and Collis, L. (1993) *Aggregates*, Geological Society Engineering Geology Special Publication No. 9, 399pp.

Stack, C.E. and Schnake, M.A. (1983) Refractory clays, in *Minerals in the Refractories Industry – Assessing the Decade Ahead* (eds P.W. Harben and E.M. Dickson), *Industrial Minerals*, **187** (April 1983), 69–77.

Strauss, G.K., Madel, J. and Alonso, F. (1977) Exploration practice for stratabound volcanogenic sulphide deposits in the Spanish–Portuguese pyrite belt: geology, geophysics and geochemistry, in *Time- and Strata-bound Ore Deposits* (eds D.D. Klemm and H.-J. Schneider), Springer Verlag, Berlin, 55–93.

Strauss, G.K., Roger, G., Lecolle, M. and Lopera, E. (1981) Geochemical and geologic study of the volcanic-sedimentary sulfide body of La Zarza, Huelva Province, Spain. *Economic Geology*, **76**, 1975–2000.

Swaddle, T.W. (1990) *Applied Inorganic Chemistry*, University of Calgary Press, Calgary, 331pp.

Taylor, J.C.M. (1984) Late Permian-Zechstein, in *Introduction to the Petroleum Geology of the North Sea* (ed. K.W. Glennie), Blackwell, Oxford, 61–83.

Velde, B. (1992) *Introduction to Clay Minerals*, Chapman & Hall, London, 198pp.

Vincent, A. (1983) The origin and occurrence of Devon ball clays, in *Residual Deposits: Surface Related Weathering Processes and Methods* (ed. R.C.L. Wilson), Geological Society of London, Special Publication No. 11, 39–45.

West, D.R.F. (1982) *Ternary Equilibrium Diagrams*, 2nd edn, Chapman & Hall, London, 149pp.

Whitbread, M. and Marsay, A. (1992) *Coastal Superquarries to Supply South-east England Aggregates Requirements*, Department of the Environment Geological and Minerals Planning Research Programme, Her Majesty's Stationery Office, London, 49pp.

Woods, P.J.E. (1979) The geology of Boulby Mine. *Economic Geology*, **74**, 409–18.

Index

AAV, see Aggregate abrasion value
Abbrook Member (Bovey Basin) 59
Absorptive capacity 237
Acetogenesis 234
Acid rain 92
ACR, see Alkali–carbonate reactivity
ACV, see Aggregate crushing value
African Rift Valley 75–6
Aggregate abrasion value 23–4
Aggregate crushing value 23–5
Aggregate density 26
Aggregate impact value 22–6
Aggregate reactivity 27–30
Aggregates (general) 6–8
Air pollution 227
AIV, see Aggregate impact value
Albania 196
Alberta tar sands 91
Albite 47
Alite 144–9
Alkali–carbonate reactivity 29
Alkali–silica reactivity 27–30, 151, Plate B(i)
Alsace (France) 77
Alumina
  prices 4
  refractories 186
Aluminosilicate refractories 190–4
Alunite 45
Amblygonite 133
Ammonia 72–4, 232
Anaerobic conditions 231
Analcite 93, 94
Anatase 39
Anatexis 138
Andusalite
  prices 5
  refractories 191, 194
Anhydrite 76, 158

Anorthite 174
Apatite 83–8
Aplite
  prices 4
  use in glass industry 121
Archaeological sites 228
Argentina 82
Armargosite 66
Arsenic 82
Artificial liner 235
Asbestos 4, 14
Aspdin, Joseph 141
ASR, see Alkali–silica reactivity
ASTM C88 24
Attapulgite 66
Attenuate and disperse sites 236–7
Attenuation 235
Austria 138, 195

Bacteria 230
  methanogenic 233, 234
Bagshot Beds 50
Ball clay
  in brick manufacture 175
  prices 4
  processing 60–3; Plate A(iv)
Baritite 155
Baryte 3, 4, 155
Basic oxygen furnaces 189
Basic slag 187–90
Bassanite 147, 155, 166
Batch kiln 167
Bauxite 2, 4, 155
  refractories 186, 191, 194
Beidellite 64
Belite 144–90
Bentonite
  evaluation 69
  in situ deposits 66–7

INDEX

landfill site liners 229, 243
prices 4
sedimentary deposits 67–9, 131
terminology 65–6
uses 69–71
Bentonite-enriched soil 243–4
Bentonization 67
Bikita (Zimbabwe) 138
Biochemical oxygen demand 232
Biotite 42, 50
Bischofite 76
Blast furnace refractories 187
Blast furnace slag
aggregate 32
cement additive 147
conditioning 198–9
in glass manufacture 122
Blasting 227
Bloating 166
Blödite 76
BOD, see Biochemical oxygen demand
Bodmin Moor (England) 50
Boehmite 39, 161
Boracite 83
Borax 5, 83
Borehole data 201–3
Borehole database 209
Borehole posting map 206, 208
Borehole profiles 206, 207
Boron 5, 82
Boron nitride refractories 186
Boron (USA) 82
Boulby mine 77, 78
Boulder Clay 166
Boundary curve 106
Bovey Basin 50, 55–9
Bowen N.L. 98
Brick clay compositions 166, 179–83
Brick colour 174–5
Brick firing 164, 166–75
Brick manufacture, kiln design 167–8
Brick mineralogy 175–6
Bricks, handthrown 168
Brickworks 159–63
Brightness 35
Brines, wild 81
Britain, mineral production 9, 10–12
British Geological Survey 16
Brushite 84
BS 812: 1985 22
BS 812: 1989 23, 24
BS 812: 1990 22

BS 882: 1992 26
Building Research Establishment 30, 34
Bulgaria 96
Bull's eye error 209–11
Buller's rings 167
Bunter Pebble Beds, see Sherwood Sandstone Group
Bushveld complex (South Africa) 196
Buxton (England) 31, 74, 130

C_2S 144–9
C_3A 1449–9
C_3S 144–9
C_4AF 144–9
Calcining 115–18, 142–6, 155, 190
Calcite 161
Calcium bentonite 39, 66
calcium carbonate 5
Calcium montmorillonite 66
Canada 156
Carbohydrates 230
Carbon dioxide 232
Carbonate–fluor apatite 83
Carbonatite 86
Carnallite 76
Cassiterite 2, 3
Cat litter 8, 35
Cation exchange capacity 69–71
see also Ion exchange
Caustic soda 73, 77
Cement factories, location 154–5
Cement
high alumina 150
low alkali 150–1
quick setting 150
sulphate resistant 150
see also Portland cement
Cemeteries 204, 228
Ceramic clay 36, 55, 59
Ceramic tiles 35
Chabazite 92, 94–6
Chalk 31
Chamotte
prices 5
refractories 191, 192
Channel Tunnel 29
Charnwood Forest (England) 17, 19, 31
Chelford Formation 128–30
Chemical oxygen demand 232
Chert 131
Cheshire (England) 81, 128
Chile 82, 92

INDEX

China 43, 82
China clay, see Kaolin
Chlorine 73, 77, 184
Chlorite 161, 165
Chrome refractories 186
Chromite 2, 5, 127, 195–6
  podiform 195–6
  refractories 186, 195
CIS (Soviet Union) 82
Claus process 91
Clay barrier 235–6
Clay fraction 39
Clay liners 241–4
Clay, refractory 5
Clee Hill (England) 17, 19, 31
Clinker, cement 142–7
Clinochlore 161, 170, 173
Clinoptilolite 96
Coal 2, 3
Coated aggregate 26
Coating clay 35, 55
COD, see Chemical oxygen demand
Codisposal 237, 239
Coke breeze 167
Coke furnace refractories 187
Colemanite 5, 83
Compatibility triangle 108
Computer hardware 260–1
Concrete 26–30
Construction industry 10–12, 13
Contact metamorphism 132
Containment sites 235–6
Contour map 206, 209–11
Cordierite 174
Corundum 191
Cover, waste disposal sites 240
Cristobalite 67, 100–5, 174
Cuba 96
Cullet 140

Darcy's law 237
Dartmoor (England) 50
Data gridding 209–20
Datamine 200
Dawsonite 74
Death Valley (USA) 82
Decarbonation 115–17, 140
Department of the Environment 229
Desulfovibrio desulfricans 89
Detergent 14, 15, 92
Devon (England) 138
Diagenesis 44–5
Diaspore 39, 161

Diatomaceous earth 196
Diatomite 5, 196
Dickite 40, 43, 46
Digital terrain model 203–4
Dilute and disperse sites 236–8
Dolerite, aggregate 23, 26, 31–3
Dolime 190
Dolomite
  aggregate 17, 31
  in cement raw materials 152
  in glass raw materials 122, 125, 131
  refractories 190
Dolomitization 131
Dorset (England) 40, 41, 50
Drill hole data 201–3
DTM, see Digital terrain model
Durham (England) 31

East Midlands (England) 157–8
Efflorescence 166
Engineered bentonite 40, 66, 244
Enstatite 102–5, 187
Epsomite 76
Erionite 93, 94–6
Etruria Marl 166
Ettringite 149
Eucryptite 139
Eutectic 102
Evaporation/transpiration 237
Evaporites 72–82

Faujasite 92
Feldspar
  glass industry 121
  prices 5
Ferrierite 93, 94
Fertilizer 72, 73, 83, 88
Field boundary 101
Filler clay 35
Finland 86
Fireclay 3, 166, 175
  refractories 192–3
Flashing 174
Flint 131
  aggregate 23
  clay 192
Float glass process 120, 122–3
Flue gas
  desulphurization 92, 158
  emissions 166
Fluid inclusions 47–8
Fluoride glass 121
Fluorine 83, 88, 131, 184

INDEX

Fluorite 3, 5, 153
Fluorspar 5
Fontainebleau (France) 128
Forsterite 102–5, 174, 187
   refractories 186
Foundry sand 5, 128, 187
France 138, 157
Francolite 83, 88
Frasch process 89
Fuller's earth 4, 8, 39, 66, 68

Gabbro, aggregate 23
Galena 2, 3
Gannister 187
Gas flare 241
Gas migration 241
Gawsworth Formation 128
Gaza 88
Georgia (USA) 40, 42, 49, 60
Geostatistics 200
Geotextile 229
Geotextile–bentonite quilts 236, 244
Germany 96
Gibbsite 39, 150, 161
Gismondine 94
Glaserite 76
Glass
   colour 121–2
   grade spodumene 4
   sand 5
Glensanda (Scotland) 17, 20, 31
Gmelinite 93
Goethite 39, 161
Goosehams Member (Bovey Basin) 53
Granite 50
   aggregate 23, 26, 31, 32
Graphite 5
Gravel, aggregate 17, 23, 30–3
Great Sulphur Vein 90
Greece 67, 196
Green River Formation 74–5
Greenbushes (Australia) 138
Greisening 47
Groundwater 88, 228
   infiltration 237
Guano 84
Gypsum
   in brick manufacture 161, 166
   in cement 147, 153
   in evaporite deposits 76
   price 4

Haematite 156, 161, 174, 175

Halite 76–82
Halloysite 38, 40, 44
Halokinesis 77
Harding pegmatite 139
Harmotome 93, 94, 96, 97
Hazardous and Solid Waste Amendment 229
HDPE, *see* High density polyethylene
Hectorite 64, 66
Heritage 204, 228
Heulandite 93, 94, 95, 96
Hexahydrite 76
High alumina clay refractories 186
High density polyethylene 236
Hingston Down (England) 31
Hornfels, aggregate 23
Howlite 83
Hungary 96
Hydroboracite 83
Hydrocarbons 164
Hydrogarnet 148–9
Hydrogen isotopes 48
Hydrogen sulphide 3, 89–91
Hydrothermal alteration 45, 131
Hydrothermal mineralization 52
Hydroxy-fluor apatite 83

Iceland 96
IDS, *see* Inverse distance squared method
Igneous rock, aggregate 17, 31, 34
Ijolite 86
Illite
   in brick clays 161, 165, 170, 173
   composition 193, 194
   mineralogy 36–9
   in refractory clays 193
   specific surface area 69
   stability 65
Ilmenite 174
Interlayer site 36–9
Inverse distance squared method 212
Inyoite 83
Ion exchange 36, 69–71
Iraq 90
Ireland 138
Iron oxide
   pigments, prices 4
   waste 155
Isothermal section 111–12
Italy 67, 96

Japan 67, 92, 96

INDEX

Join 107
Joints 48

Kainite 76
Kao Lin (China) 43
Kaolin
 brightness 201–26, 249, 250
 computer evaluation 200–26
 evaluation: reserve quality tables 247–54
 grittiness 201–26, 254
 particle size distribution 60, 62
 prices 4
 primary deposits 39, 42–8
 processing 60–3; Plate A(i–iii)
 in refractory clays 191
 rheology 60, 201–26, 251, 252
 secondary deposits 39, 48–9
 south-west England deposits 49–63
 specifications 57
 TiO_2 content 201–26, 253
 viscosity 57, 60, 201–26, 252
Kaolinite
 in brick clays 161, 165
 composition 192, 193
 firing behaviour 170, 172, 173
 mineralogy 36–63
 specific surface area 69
 stability 45–7, 65
Kaolinization 43–8
Kazakhstan 196
Kenya 75–6
Kernite 83
Keuper Marl, see Merica Mudstone Group
Khibiny (Russia) 86
Kieselguhr 196
Kieserite 76
Kiln linings
 burning zone 196
 calcining zone 196
Kurnakovite 83
Kyanite
 price 5
 refractories 194

Lake Gosiute 74–5
Lake Magadi 75–6
Land's End (England) 50
Landfill gas 184, 232–5
Landfill leachate
 collection 237–41
 composition 231–3
 monitoring 237–41
Landfill waste disposal 228–44
 site lining materials 35, 241–4
Landscape 227
Langbeinite 76
Laumontite 94–5
Lead crystal glass 121
Leonite 76
Lepidocrocite 161
Leucite 174
Lever rule 104
Levyne 93
Lime, hard burnt 143
Lime saturation factor 151–2
Limestone
 aggregate 17, 23, 26, 31, 32
 in bricks 161, 164
 raw material for cement manufacture 146, 151–4
 raw material for chemical industry 72–4
 raw material for glass industry 122, 125, 130–6
 use in flue gas desulphurization 92
Limonite 161
Lipids 230
Liquid immiscibility 105
Liquid waste 237
Liquidus 102
Lithium 132
Lithium carbonate 4, 132
Lithium mica 50
Lithium montmorillonite 66
Lithological code 203
Loch Aline (Scotland) 127
LOI, see Loss on ignition
Lorraine (France) 74
Los Angeles abrasion value 24
Loss on ignition 113
Loughlinite 66
Löwite 76
LSF, see Lime saturation factor
Luxulyan (England) 31

Macadam, James 21
MacOreReserve 200
Magnesia
 dead burned 190
 see also Periclase
Magnesian Limestone 130, 190
Magnesite 4, 117–18, 190
 refractories 186

INDEX

Magnesium montmorillonite 66
Magnesium sulphate soundness value 24
Magnetite 174, 175
Marcasite 39
Marine aggregate 28
Marine gravel 23
Matrix 39
Meerschaum 66
Mendip Hills (England) 17, 19
Mercia Mudstone 166
Mercia Mudstone Group 157–8
Mesolite 94
Metabentonite 66
Methane 3, 232–5
Methanogenesis 234
Meyerhoffite 83
Mica 4
Microbiological activity 230–1
Micro-Deval value 24
Midland Valley (Scotland) 31
Mine planning 226
Mine waste 229
Mineral
  definitions 1–2
  markets 1
  prices 4–5
  production and politics 12–14
  production information 14–16
  workings
    impact on rights of way 228
    impact on residents 228
    impact on tourists 228
    impact on water quality 228
Mineral Resources Consultative Committee 16, 34
Mississippi bentonite 66
Mississippi Valley Type mineral deposits 2, 3, 131
Molochite 192
Monetite 84
Montmartre (France) 155
Montmorillonite 38, 64–71
Mordenite 93, 94–6
Morocco 84, 87
Moulding sand 187
MSSV, *see* Magnesium sulphate soundness value
Müller-Kühne process 92
Mullite 174, 191, 194
  ore 194
  refractories 186

Mundic 28
Muriate of potash 77
Muscovite 47

Nacrite 40, 43
Nahcolite 74
Namibia 84, 138
Native sulphur 89–90
Natrolite 94–5
Natron 74
Natural gas 91
Nepheline 86
Nepheline syenite
  prices 4
  raw material for glass industry 121
New Zealand 96
Newark Gypsum 158
Non-swelling bentonite 66
Nontronite 64
North Carolina (USA) 138
Northumberland (England) 31
Norway 19, 195
Notation, ceramic 118–19
Nottinghamshire (England) 158

Oilfield drilling fluids 35
Olivine 5, 14, 195, 196
Olivine sand 195
Oolite, aggregate 32
Optical fibres 121
Orefields
  Pennine 9, 90
  south-west England 9
Orpiment 82
Orthoclase 47, 174
Outer Hebrides (Scotland) 19
Overburden 203
Oxford Clay 166
Oxygen
  fugacity 174, 175
  isotopes 48

Paint 35
Palabora (South Africa) 86
Palygorskite 66
Paper 35
Paragonite 47
Paris Basin 74
Partial melting 143
Particle size distribution 60
PC/Cores 201
Peak District (England) 132–6

INDEX

Pegmatite 134
Periclase 102–5, 153, 187, 189–90
Peritectic 105
Perlite 155
Permo-Trias, Europe 157
Peru 82
Petalite 4, 133, 138, 139
Petrockstowe Basin 50, 59
Petroleum 91
PFA, see Pulverized fuel ash
Pharmaceuticals 35
Phase diagram (general) 98–116
phase rule 99–112
Philippines 196
Phillipsite 92, 94
Phosphates 5, 14–15
Phosphoria Formation 85
Phosphoric acid 83
Picromerite 76
Place value 6
Planning application 227
Planning permission 227
Plaster of Paris 155
Plasterboard 155
Plastics 35
Poland 90
Polished stone value 23–6
Polyhalite 76
Portland cement
  manufacture 141–7
  setting 147–9
Portlandite 144, 147–9
Potash 5, 78
Potassium bentonite 66
Potassium montmorillonite 66
Pozzolan 147
Priceite 83
Probertite 83
Protein 230
PSV, see Polished stone value
Pulverized fuel ash 27, 147
Putrescible waste 229–31
Pyrex 121, 132
Pyrite
  in clays 39, 161, 166
  source of sulphur 3, 90–1
Pyrite Belt (southern Iberia) 90
Pryophyllite 46, 64, 65

Quartz 100–5, 174
Quartz sand refractories 189
Quartzite, aggregate 23

Railway ballast 34
Rainfall infiltration 237
Rare-metal pegmatite 134
Reaction point 105
Realgar 82
Refractory bricks 185
Refractory clays, composition 193
Refractory grog 192
Reserve evaluation software 260
Reserve modelling 220
Resource Conservation and Recovery Act 229
Road construction 21–6
Road traffic 227
Rutile 2

Salt (halite) 5
Salt cake 77
Salt dome cap rock 89
Sand
  aggregate 17, 30–3
  raw material for glass industry 122, 126–30
Sandstone, aggregate 23, 26, 33
Sanitary waste disposal, see Landfill waste disposal
Sanitaryware 35, 36
Saponite 64, 66
Sassolite 83
Sauconite 64
Schonite 76
Searles Lake (USA) 75, 82
Seawater 190
Secondary aggregate 26
Sepiolite 66
Serpentine 195, 196
Severn estuary (England) 31
Shale
  raw material for brick manufacture 166
  raw material for cement 146, 152–4
Shap (England) 31, 32
Sherwood Sandstone Group 30
Sicily 90
Siderite 39, 161
Sierra Nevada (USA) 75
Silica fume 147
Silica
  polymorphs 100–5, 169, 186–7
  refractories 186–9
  sand 5, 126–30
Silicon carbide refractories 186
Silicosis 14

INDEX

Sillimanite 5
  refractories 186, 191, 194
Site of special scientific interest 228
Slag, *see* Blast furnace slag
Slake durability test 24
Slate 5
Smectite
  in bentonites 63–71
  in brick clays 161, 165, 170, 173
  mineralogy 36–40
Smectite to illite reaction 68
Smelter gas 90
Soda ash 5
Soda–lime–silica glass 120
Sodalite 93
Sodium bentonite 39, 66, 244
Sodium montmorillonite 66
Sodium sepiolite 66
Sodium-activated bentonite 66
Sodium-exchanged bentonite 66
Solfatara 45
Solidus 102
Solvay process 72–4
Solvus 105
Sour gas 91
South Africa 96
South Carolina (USA) 40, 42
Southacre Member (Bovey Basin) 56, 59
Southern bentonite 66
Spain 19, 138, 156, 195
  mineral production 9
Sphalerite 2, 3
Spinel 174
  refractories 186
Spodumene 4, 133, 138, 139
Spodumene-quartz intergrowth 138
Squi 138
SSSI, *see* Site of special scientific interest
St Austell (England) 49–52, 55
Stability diagram
  activity–activity 64
  activity–temperature 46, 65
Stanley Bank Basin 50
Steel, austenitic 195
Steelmaking 189
Sticklepath–Lustleigh fault 50, 59
Stilbite 94
Stover Member (Bovey Basin) 56, 59
Sub-bentonite 66
Sulphate soundness test 24
Sulphide ore minerals 90–1

Sulphur
  bright 88
  dark 88
  native 89–90
  price 5
  slate 89
  volcanic 91
Sulphur dioxide 91–2, 184
Sulphuric acid 73, 89–92
Superphosphate 83
Superquarries 8, 17
Surface area 69
Surface water infiltration 237
Swelling bentonite 66
Sylvite 76
Syngenite 76
Synthetic bentonite 66
System
  definition 98
  one-component 100–1
  two-component 101–5
  three-component 105–12
  SiO_2 100–1
  SiO_2–Al_2O_3 190–1
  SiO_2–MgO 102–5, 187–90
  SiO_2–Al_2O_3–CaO 144–6, 151, 170, 173, 178–80, 246
  SiO_2–Al_2O_3–FeO 178–80, 248
  SiO_2–Al_2O_3–K_2O 170, 173, 178–82, 249
  SiO_2–Al_2O_3–MgO 113–14, 170, 173, 178–80, 250
  SiO_2–CaO–Na_2O 123–5
  SiO_2–MgO–CaO 106–12, 198–9, 247
  CaO–Al_2O_3–Fe_2O_3–SiO_2 151
Szaibelyite 83

Talc
  prices 5
  mineralogy 64
Tanco (Canada) 138
Tarmac 21–6
Tarmacadam 21
Ten per cent fines value 22, 26
Ternary eutectic 110
Texas bentonite 66
Thailand
  mineral production 9–10
  plaster production 156
Thames Valley (England) 30
Thermonatrite 74
Thomsonite 94

INDEX

Tie line 107
Time–temperature–transformation diagram 116–18
Tincal 83
Tincalconite 83
Tourmaline 42, 50
Transport Research Laboratory 34
Trent valley (England) 30
Tridymite 100–5, 174
Triple point 101
Trona 74
TTT diagram, *see* Time–temperature–transformation diagram
Tunnel kiln 167–8
Turkey 82, 196
Tutbury Gypsum 158

Ulexite 83
United States Bureau of Mines 16
USA 96, 195

Vale of Eden (England) 157–8
Vanthoffite 76
Vermiculite 5, 155
VFA, *see* Volatile fatty acid
Virgilite 139
Volatile fatty acid 231–2
Volcanic ash 67–8, 96, 131
Vycor 121

Wairakite 94
Washington degradation test 24
Waste disposal 184
Waste Management Papers 229
Waste solvents 143
Waste stabilization 231
Waste tyres 143
Water balance 237
Weathering 43
Weathering index 24
Wegscheiderite 74
Western Canada Basin 77, 91
Westerwald (Germany) 40
Whin Still 31
Whiteness 35
Wollastonite 5
Wyoming (USA) 74–5
Wyoming bentonite 39, 66, 244

X-ray fluorescence 130, 151, 164, 177

Zechstein province 77
Zeolite A 93
Zeolites 14, 93–6
Zircon 5, 195
Zirconia
   refractories 186, 194
   sand 194–5